W9-CMJ-446
R0174386144
The universe unveiled : instruments

The Universe Unveiled

The Universe Unveiled

Instruments and Images through History

Bruce Stephenson
Marvin Bolt
Anna Felicity Friedman

E
T
25
30
35
40
45
50
55
60
65
70
75
80
85
90
D
B
12

Acknowledgments

We hope that readers of all backgrounds will appreciate this book for the inspiring story carried in its images: the story of how we came to know so much about our world. Instruments of science, images of nature, snapshots of humankind's investigation of the universe—and the vehicles that spread the resulting knowledge to unprecedented numbers of people—have been crucial in building our current body of scientific knowledge. A multitude of individuals—some famous, some unknown—contributed to this knowledge while striving to understand the world they lived in. Our first acknowledgment must be to them. We created this book to bring their story to a wider public.

Ed Marquand of Marquand Books coordinated the design and production of this book, providing us with astute insights into the process of publishing. We also thank the staff of Marquand Books, especially Vivian Larkins, Marie Weiler, and John Hubbard, and proofreader Laura Iwasaki. Suzanne Kotz copyedited the text, and we genuinely appreciate her perceptive attention to detail.

Steve Pitkin and Catherine Gass took the photographs, almost all of which are new. Devon Pyle-Vowles managed the logistics of our studying and photographing several hundred delicate artifacts and rare books. Patti Gibbons and Jennifer Brand assisted her. Roy A. Kaelin Jr. produced the drawings for the Astronomical Appendix. Mave Lawler cheerfully helped in every way we could ask, sparing us a great deal of clerical work. Linda Chinery organized and labeled the photographs, compiled the index, and assisted in many other ways.

We are grateful to all the staff at Cambridge University Press who worked on our project and, in particular, to our editors Alice Houston and Adam Black.

Bruce Chandler, Owen Gingerich, Bob Karrow, Liba Taub, and Marjorie Webster read early drafts of the text and brought to our attention many infelicities and a few outright errors. We thank them for their willingness to help us. For any errors that remain, we take full responsibility.

Paul Knappenberger, President of the Adler Planetarium & Astronomy Museum, has consistently and strongly supported the use of history in presenting the human side of astronomy to the public. We hope that this book contributes to the new Adler he is building for the 21st century.

We thank our friends, families, and all the Adler staff for providing support and encouragement for the project.

The three authors divided the writing equally, but one of us, Anna Felicity Friedman, was both moving spirit and project manager for this book. She coordinated our dealings with Marquand Books and Cambridge University Press; she maintained our book maps and photography notebook; and above all, she prodded her coauthors to produce their pages on schedule. Without Anna's efforts we might never have started this book, and we certainly would never have finished it.

Marjorie Webster and the late Roderick Webster, longtime curators at the Adler, are more than anyone else responsible for the present strength of its History of Astronomy collections and the existence of its History of Astronomy department. We dedicate this book to them.

Bruce Stephenson
Marvin Bolt
Anna Felicity Friedman
Chicago, June 2000

MISPHÆRI
OREALE
SPHÆRIO
UM STI
CVM SU
TERI
AMERICA
Nova Brit
SEPTENTRIO
tannia
NALIS
Buttons bay
Draco
Terra Nova
Cepheus
Ursa minor
Andromeda
Caßio peia
Tropicus
Bootes
Coma Berenices
Cancri
VIRGO
OCEANVS CHINENSIS
Perseus
Caput Medusæ
ARIES
TARTARIA
Lordanis fluv
GEMINI
CANCER
LEO
Zodi
Borneo
Dagali
Canis minor
Æquinoctialis
Cor Hydræ
Monoceros
Unicornis
Lepus
MARE

Introduction

The instruments and images depicted on the following pages encapsulate a specific part of the history of the unveiling of the universe, roughly from the mid-1400s to the end of the 1800s, focusing primarily on the Western world but also touching on Islamic and East Asian contributions. This period witnessed great invention and change, including possibly the most significant and counter-intuitive philosophical shift ever—the change from an Earth-centered universe to a Sun-centered one.

The instruments that early scholars employed—armillary spheres, astrolabes, globes, sundials, telescopes— arouse our curiosity. Rare books and atlases containing celestial charts, astronomical diagrams, and other illustrations likewise captivate us. Often mysterious and always impressive, these artifacts variously intrigue us with their function and construction, amaze us by their age, and enthrall us with their beauty. We want to know who made them and why. We want to learn the stories they have to tell.

Today such objects grace the shelves of public and private collections around the world. They function as symbols of an enduring quest for knowledge about our world and of the discoveries that have enabled us to expand our understanding of the universe. They stand not only as testaments to the ingenuity and intelligence of their makers, but as relics of changing worldviews. In manifesting the skills of artisans now long forgotten and in representing social orders that have passed away, such artifacts afford a glimpse into the world of our ancestors.

Creating Scientific Instruments and Images

From before the dawn of history, people have observed and interpreted the heavens in their quest to survive, to live in communities, and to find meaning. The compelling events playing

Instrument Buch durch Petrum Apianum erst von new beschriben.

Zum Ersten ist darinne begriffen ein newer Quadrant / dardurch Tag vnd Nacht / bey der Sonnen / Mon / vnnd andern Planeten / auch durch ettliche Gestirn / die Stunden / vnd ander nutzung / gefunden werden.

Zum Andern / wie man die höch der Thürn / vnd anderer gebew / des gleichen die weyt / brayt / vnd tieffe durch die Spigel vnd Instrument / messen soll.

Zum Dritten / wie man das wasser absehen oder abwegen soll / ob man das in ein Schloß oder Statt füeren möge / vnd wie man die Brünne suchen soll.

Zum Vierden / findt drey Instrument / die mögen in der gantzen welt bey Tag vnd bey Nacht gebraucht werden: vnnd haben gar vil vnd manicherlay breüche / vnd alle geschlecht der Stunden / behalten alle zu gleich jre Lateinische nämen.

Zum Fünfften / wie man künstlich durch die Finger der Hände die Stund in der Nacht / on alle Instrument erkhennen soll.

Zum Letzten / ist dariñ ein newer Meßstab / des gleichen man nendt den Jacobs stab / dardurch auch die höch / brayt / weyt / vnd tieffe / auff newe art gefunden wirt.

INGOLSTADII Cum Gratia & Priuilegio Cæsareo ad Triginta Annos. AN. M.D.XXXIII.

out on the celestial stage inspired the human imagination to create an astonishing array of stories and myths. Astronomical lore passed down from generation to generation in various forms: via oral traditions and ceremonies; as notches on bones, sticks, or rocks; in elaborate structures; and sometimes in written form.

However elaborate or simple the knowledge embodied in these forms, it usually proved sufficient to meet the needs of a community. In a few cultures, that knowledge took on a more abstract character, as people recorded their observations, calculations, hypotheses, and theories in manuscripts. These writings were copied throughout the ancient world, the Near and Far East, and, later, in medieval and Renaissance Europe. Astronomical and mathematical illustrations in such texts reproduced what observers saw or helped readers visualize abstract concepts. Decorative illuminations evoked reverence for the information contained within. Some ancient Greek texts described devices or their mathematical foundation, including the stereographic projection later used in astrolabes and star charts. The oldest surviving complex instrument is the Antikythera mechanism, a device constructed in the Mediterranean region in the first century B.C. to calculate the relative positions of the Sun and the Moon.

Most of the scientific instruments we see today date from the Renaissance or later, though a few still survive from the turn of the first millennium or even earlier. Artisans (some of them astronomers and philosophers) crafted them to aid their investigations or to demonstrate and display recent discoveries. Many instrument makers also wrote books that functioned as owner's manuals, served as less expensive substitutes for instruments, or gave directions for making an instrument.

In the 1600s and 1700s, books and instruments were no longer the exclusive property of the elite or the scholar. Books such as Apian's *Cosmographia* functioned as astronomy or geography textbooks for those who could obtain and read them. Because of their cost, celestial and terrestrial atlases appealed to the wealthy middle class, for whom astronomical knowledge increasingly constituted a status symbol. More didactic volumes included paper instruments, such as volvelles, or offered instruction in the use of actual instruments. Instruments such as pocket globes for teaching geography to children also aided in popular education.

The popularity of astronomy in recent centuries is manifest in the elaborate decoration and fine materials devoted to ornate telescopes. They were likely parlor novelties or assertions of expertise rather than instruments that anyone routinely looked through. In the 1700s, public lectures complemented the private study of the sky. Lecturers often presented the regular motions of planets as evidence of a divinely created order. To demonstrate the point, they used elaborate models of the heavens, called orreries, which they themselves might have built. In the early 1800s, the increasing scientific literacy of the general public contributed to a widening split between popular and professional audiences.

Beyond their immediate use by scholars, teachers, and students, scientific instruments and images facilitated dialogue between different cultures and disseminated knowledge. In the Islamic world, Greek texts provided the foundation for a

sophisticated astronomy. This knowledge then spread through India to China, where certain elements fused with traditional Asian astronomy. Magnetized compass needles came to Europe from China; with them, European navigators could set out on the open ocean to explore new worlds. In turn, Jesuit missionaries brought European astronomical and geographical scholarship to Chinese astronomers, who incorporated aspects of it into their concepts about the universe.

Technical developments also contributed to the understanding of the nature and structure of the universe. The invention of printing with movable type in Europe about 1450 eliminated the tedious and expensive process of copying manuscripts and allowed for multiple copies of works to be distributed more widely. The refinement of measurement techniques, especially those involving the engraving of numerical scales, yielded instruments that could detect small, subtle changes in the positions of heavenly objects with far more precision than ever before. With the development of advanced metalworking and woodworking techniques, mechanisms for clocks and other devices gained in accuracy but also in complexity. Just as significant for astronomy was the discovery of spectroscopy, the analysis of light from heavenly objects to identify their physical properties.

Artistic developments, such as improvements in perspective drawing and the printing of images, allowed increasingly detailed visualizations of ideas about the universe. Copperplate engraving, developed in the 1500s, largely replaced cruder woodcut prints and allowed greater artistry and intricacy. Celestial charts became increasingly adorned, often with hand-coloring. Later printing processes, including lithography, invented in 1798, and the very fine steel engraving of the 1800s, continued to improve the representational quality of scientific illustrations. Photographic printing processes, developed in the mid- to late 1800s, finally removed the hand of the artist from the rendering process, enabling astronomers to provide accurate illustrations of such subjects as nebulae and planetary surfaces.

Instruments and Images in Art and as Art

The historic instruments and books we admire today once represented the knowledge and prestige of their owners. They often appeared in paintings, prints, and other illustrations to evoke symbolic or allegorical meanings. People in Europe, the Middle East, and India regarded armillary spheres and

celestial and terrestrial globe pairs as symbols of the universe, astronomy, and even astronomers long after the widespread acceptance of a Sun-centered model of the universe. Armillary spheres and globes, often accompanied by human figures, decorate the pages of atlases and texts about astronomy, cosmography, geography, and navigation.

In historical portraits, the astrolabe, the all-purpose tool of so many astronomical functions and calculations, often identified a person as an astronomer. Later, the telescope served the same identifying function. Astronomical instruments used for sailing by the stars and specialized navigational instruments, such as cross-staffs and back-staffs, indicated navigators or explorers. Terrestrial globes typically signified cartographers or geographers.

But instruments functioned not only as symbols in art. Elaborately decorated versions were works of art themselves when crafted of fine materials like gilt brass and ivory. At times strictly aesthetic or even ostentatious, these embellishments frequently summoned the same allegories as printed matter or fine art. For example, celestial globes often are supported by stands in the shape of the mythological Atlas, who holds up the sky.

In decorations surrounding book images, artists often personified abstract scientific ideas or depicted allegorical, historical, or mythological characters. Such conceits included personifications of the planets, academic disciplines, the Muses, and famous scientists. Constellation figures not only marked regions of stars but helped people remember the patterns of the stars by relating mythological stories to them.

In the 1800s, the design of instruments and illustrations became less ornate and symbolic, reflecting a changing cultural aesthetic more in tune with an increasingly mechanized world. Form followed function, as an object or image's utility alone came to justify its existence.

VRANIA
ORONTIVS
circulus
meridianus
Vertex
Mars
Sol.
Horizon
obliquus

Deo et Posteritati
Lynceo
Multa detecta
JOHANNIS HEVELII
MACHINA COELESTIS

From Current to Future Collections

As we look at scientific artifacts today, we marvel at the pioneers who observed and studied the universe in order to make sense of it. We now take for granted the concepts they developed at great expense, time, and effort. We can only imagine the mental and physical courage needed to trust in the scientifically given knowledge that the Earth is round when embarking on the vast ocean for new lands; to envision the Earth hurtling through space, not only spinning on its own axis but orbiting around the Sun; to gaze unafraid on spectacular celestial occurrences such as comets and eclipses.

Today, our knowledge of the universe continues to expand. We still craft tools and write texts about our observations, discoveries, and theories. The museums and libraries of the future will house artifacts very different from those included in this book to represent the accomplishments of today's inventors, scholars, and scientists. Texts now can be published virtually, illustrations exist only as digital files, and artifacts are enormous or ephemeral. We cannot conceive, any more than our predecessors could, what a book such as this will resemble a few centuries in the future.

Note to the reader: In the pages that follow, words marked with a "+" are defined in the glossary on pages 140–41.

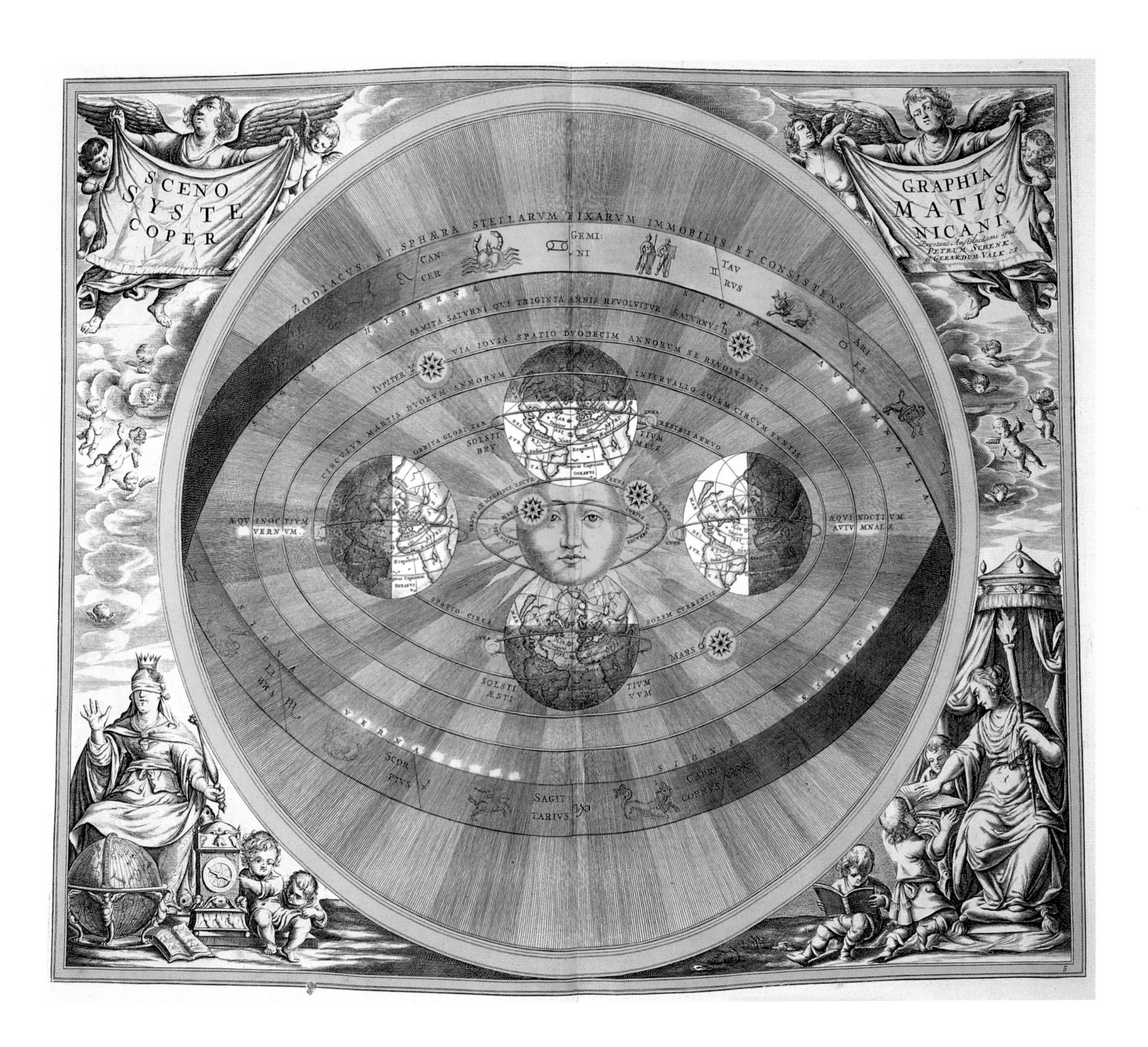
SCENO
SYSTE
COPER
GRAPHIA
MATIS
NICANI.
Prostant Amstelodami apud
PETRUM SCHENK
et GERARDUM VALK
CAN
CER
GEMI
NI
TAV
RVS
ARI
ES
SEMITA SATVRNI QUI TRIGINTA ANNIS REVOLVITUR SATVRNUS
VIA IOVIS SPATIO DVODECIM ANNORVM SE REVOLVENTIS
IVPITER
INTERVALLO SOLEM CIRCVM
CIRCVLVS MARTIS DVORVM ANNORVM
ORBITA GLOBI TER
RESTRIS ANNVO
SOLSTI
BRV
TIVM
MALE
ÆQVINOCTIVM
VERNVM
ÆQVI NOCTIVM
AVTV MNALE
SPATIO CIRCA
SOLEM CVRRENTIS
MARS
SOLSTI
ÆSTI
TIVM
VVM
SCOR
PIVS
SAGIT
TARIVS
CAPRI
CORNVS

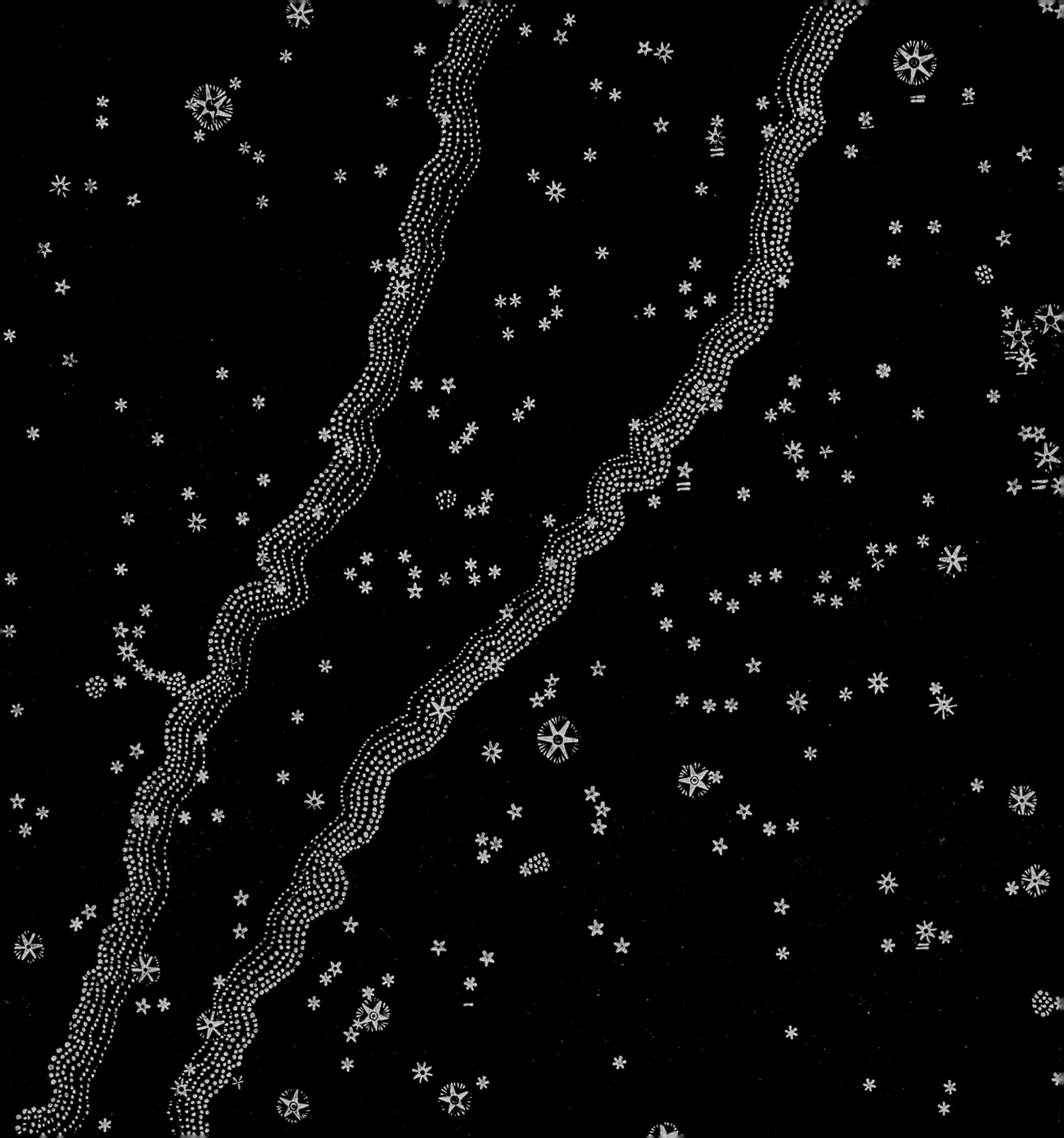

Discovering Space

Looking through clear skies at the starry night, we feel the same awe and wonder that our ancestors did long ago. We, like them, see planets move in mysterious looping paths against a backdrop of twinkling stars. These stars fill the heavens but assemble especially in the Milky Way,+ a narrow, diffuse band of light stretching across the sky.

For thousands of years, people have wondered about the cosmos+ and their place within it. To make sense of the heavens, the earliest peoples created diverse mythologies in which the behaviors of the planets, as well as the motions of the Sun, Moon, and stars, played central roles. People ordered their lives by the regularity of the patterns in the sky, but their stories also had to account for unusual objects and events such as eclipses, comets, meteor showers, and even new stars that appeared from time to time.

Increased exploration of the Earth encouraged and required more detailed information about star positions, prompting the development of new astronomical tools for navigating the seas. Over the years, navigators tinkered with those tools, improving and transforming them into more elaborate, accurate forms and into entirely new devices.

Astronomers seeking deeper knowledge of the heavens also came to rely on better instruments. After the invention of the telescope about four centuries ago, astronomers continued to focus their efforts on charting the planets and stars, and even discovered more of each. But the telescope eventually encouraged people to see the expanse of the universe in a new way: they began to think of heavenly bodies as distant but physical worlds, whose secrets the telescope would gradually unveil.

Looking at the Stars

People around the globe have looked at the night sky and seen patterns in the stars. They observed that the stars were not always present in the same place, but appeared to move in a systematic way (see appendix, p. 144 G). Noticing that the positions of the stars stayed the same with respect to one another, our ancestors named these points of light the "fixed" stars (as opposed to the "wandering" stars—the planets). Astronomers mapped the patterns of stars simply at first. As astronomical instruments became more precise, they created more complex charts.

From Lore to Order

Constellations+ organize the stars into more easily identifiable groups. No one knows the exact origin of the constellations that we use today, but 48 had been established by ancient Greek times. We call these the Ptolemaic constellations after the astronomer Ptolemy, who recorded them in the star catalogue+ portion of his book the *Almagest* [A] in the second century A.D.

Classical authors wrote about the mythology of the constellations in such works as the *Phaenomena* of Aratus [B], from the third century B.C., and the *Poeticon astronomicon* of Hyginus [E], from the second century A.D. They provided the locations of the stars within each constellation with approximate verbal descriptions. Later manuscript and printed editions of these texts often included illustrations but not accurate maps of the stars. The illustrators depicted the constellations according to the text descriptions without looking at the night sky. As a result, their images did not resemble the actual patterns in the heavens.

LIBER VII. 187

¶Expoſitio regularis conſtellationum hemiſphærij borealis.
¶Formæ boreales.

1		2		3	4	Vrſa
Nūerus	¶Minoris Vrſę conſtellatio pª	Lōgitudo G M		Latit. G M	Ma.	min.
1	Quę eſt in extremitate caudæ	♊ 0 10	bor.	66 0	3	♄ ♀
2	Quæ poſt ipſam in cauda eſt	♊ 2 30	bor.	70 0	4	
3	Quæ poſt iſtam prope radicem caudæ	♊ 16 0	bor.	74 20	4	
4	Auſtralis ſtella pręcedētis lateris figurę qua	♊ 29 40	bor.	75 40	4	
5	Borealis eiuſdem lateris (drilateræ	♋ 3 40	bor.	77 40	4	
6	Auſtralis earum quæ in ſequenti latere ſunt	♋ 17 10	bor.	72 50	2	*
7	Borealis eiuſdem lateris	♋ 26 10	bor.	74 50	2	*

¶Vrſæ minoris * 7. Magnitudinis * Secundæ 2, Tertiæ 1, Quartæ 4

Informata quę circa urſam minorem eſt.

1	Auſtraliſſima extra figurā in recta ſequētis (lateris	♋ 13 0	bor.	71 10	4	

Maioris Vrſæ conſtellatio 2ª — Vrſa maior

1	Quæ eſt in extremitate rictus	♊ 25 20	bor.	39 50	4	
2	Pręcedens earum quę in duobus oculis ſunt	♊ 25 50	bor.	43 0	5	
3	Sequens earum	♊ 26 20	bor.	43 0	5	
4	Pręcedens earum quę in fronte ſunt	♊ 26 10	bor.	47 10	5	
5	Sequens earum	♊ 27 40	bor.	47 0	5	
6	Quę in extremitate præcedentis auris eſt	♊ 28 10	bor.	50 30	5	
7	Pręcedēs earum quę in collo ſunt	♋ 0 30	bor.	43 50	4	
8	Sequens earum	♋ 2 30	bor.	44 20	4	
9	Borealior de duabus quæ in pectore ſunt	♋ 9 0	bor.	42 0	4	
10	Auſtralior ipſarum	♋ 11 0	bor.	44 0	4	mi. par.
11	Quæ in genu ſiniſtro eſt	♋ 10 40	bor.	35 0	3	39.0
12	Borealis earū quę in anterioris extremitate	♋ 5 30	bor.	29 20	3	
13	Auſtralior ipſarū (pedis ſiniſtri ſunt	♋ 6 20	bor.	28 20	3	
14	Quæ ſupra genu dextrum eſt	♋ 5 40	bor.	30 10	4	36.0
15	Quæ infra genu dextrum eſt	♋ 5 50	bor.	30 20	4	33.20
16	Earū quę ſunt in quadrilatera figura, illa in	♋ 17 40	bor.	49 0	2	* ♂
17	Quæ de iſtis in urſæ latere eſt (dorſo eſt	♋ 22 10	bor.	44 30	2	*
18	Quę in radice caudæ	♌ 3 10	bor.	51 0	3	
19	Reliqua quę eſt in poſteriori ſiniſtra coxa	♌ 4 0	bor.	46 30	2	*
20	Precedēs earū quæ in extremitate poſteriorū	♋ 22 40	bor.	29 20	3	☾ ♀
21	Quæ iſtam ſequitur (ſiniſtri pedis ſunt	♋ 24 10	bor.	28 15	3	
22	Quæ eſt in poplite ſiniſtro	♌ 1 40	bor.	35 15	4	
23	Borealiū earū que in extremitate poſterioris	♌ 9 50	bor.	25 50	3	
24	Auſtralior earum (ſiniſtri pedis ſunt	♌ 10 20	bor.	25 0	3	♌ 13.20
25	De tribus in cauda locatarū, prima poſt cau-	♌ 12 10	bor.	53 30	2	*
26	Media ipſarum dæ radicem	♌ 18 0	bor.	55 40	2	*
27	Tertia, & in ipſa extremitate caudæ	♌ 20 50	bor.	54 0	2	*

Magnitudines

A

B

C

Using Ptolemy's catalogue to locate a star proved difficult because, though it gave celestial latitude and longitude, it identified each star with a long, prose description. To describe or locate stars more simply, astronomers needed a universal nomenclature.[+]

In the 1500s, Giovanni Paolo Gallucci and Alessandro Piccolomini [D] made early attempts to standardize nomenclature. In 1603, Johann Bayer developed the most influential system, still in use today, for stars visible to the naked eye [C]. Bayer assigned letters to each star within a constellation, roughly in descending order of the brightness of the star, beginning first with the Greek alphabet and then, if all those letters were exhausted, continuing with the Roman alphabet.

D

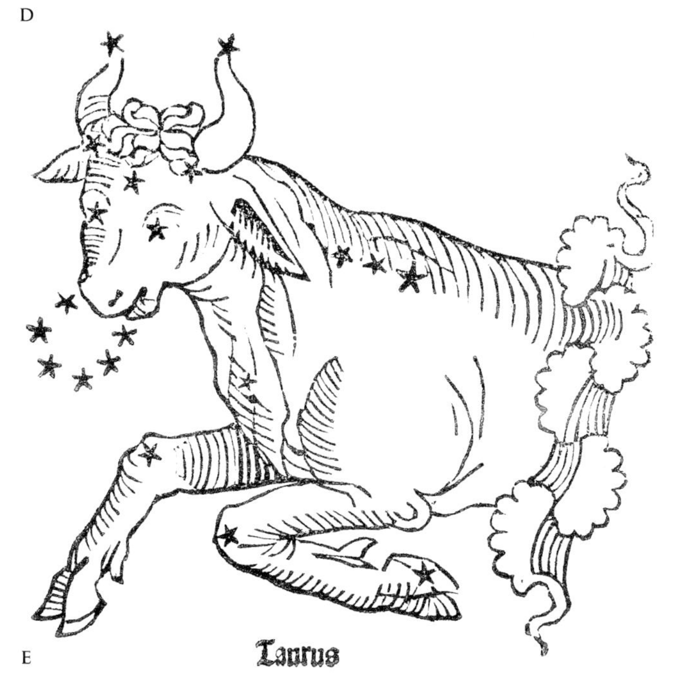

E

New Constellations

Starting in the 1500s, cartographers began to add new constellations[+] to Ptolemy's list. Caspar Vopel introduced Berenice's Hair (Coma Berenices) and Antinoüs on his globe of 1536. Many cartographers copied both of these constellations and incorporated them onto other star charts and globes [A].

In 1595, traveling as part of a Dutch expedition to the South Seas, the astronomers Pieter Dircksz Keyser and Frederick de Houtman observed the stars located around the south celestial pole.[+] Because these stars are not visible at northern latitudes,[+] early European celestial maps and globes lack constellations around the South Pole [B]. After Keyser and de Houtman returned, globe maker Petrus Plancius recorded 12 new constellations based on their findings. Several years later, Johann Bayer became the first astronomer to depict the southern skies on a map [C], which helped to popularize these constellations.

Plancius introduced several other constellations, including the Southern Cross (Crux). Although Amerigo Vespucci observed it on one of his expeditions to the New World [E] and Ptolemy had actually recorded its stars in the *Almagest*, the Southern Cross was not formally depicted as its own constellation until 1589.

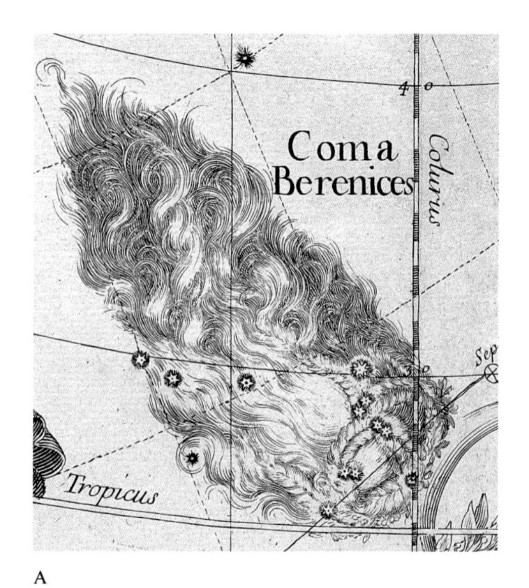

A

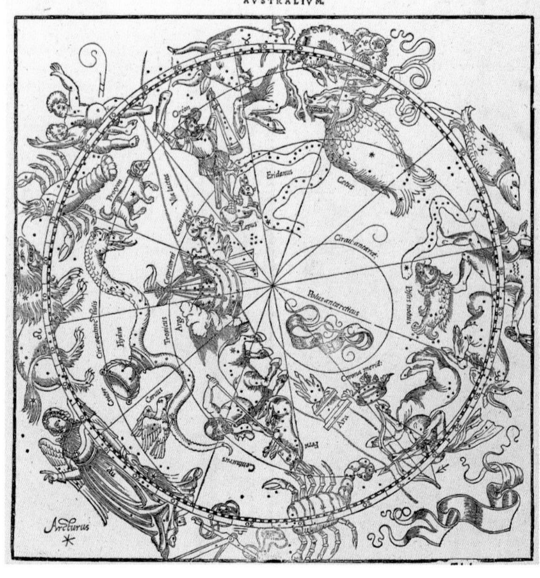

B

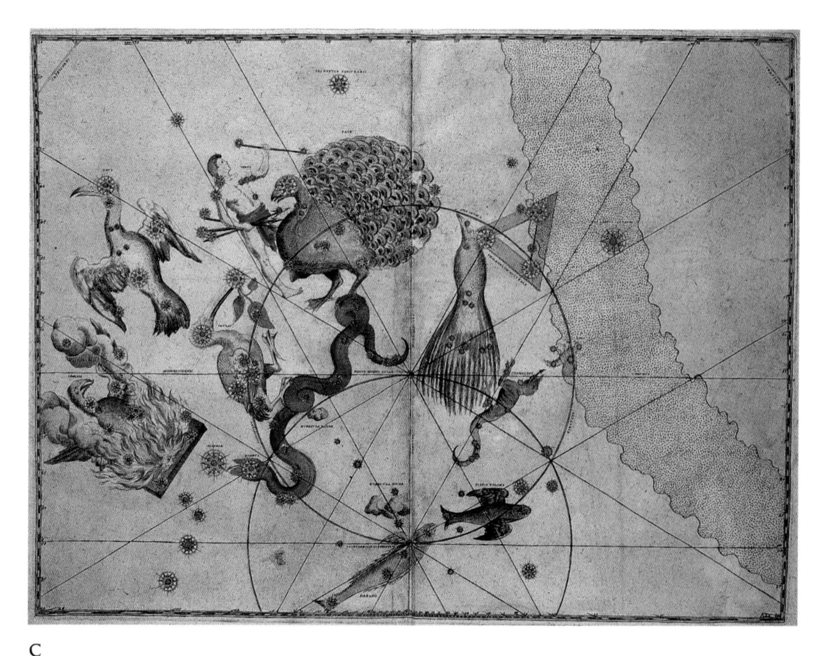

C

D

Many astronomers invented new constellations, though most have not endured. For example, in 1627, Julius Schiller proposed a reworking of the traditional mythological constellations into figures from the Bible [F].

In the late 1600s, Johannes Hevelius introduced 11 constellations, 7 of which we still use [D]. In the mid-1700s, Nicolas Louis de Lacaille contributed 14 more [G]. He also proposed splitting up the enormous Ship (Argo Navis), the only original Ptolemaic constellation not used today, and replacing it with the Keel (Carina), the Stern (Puppis), and the Sail (Vela). Today we recognize 88 official constellations.

F

E

G

Chinese Constellations

A

B

C

In ancient times, astronomers in China held high social positions. Indeed, astronomy in ancient China shows numerous close connections with the imperial court. For example, all cultures in northern latitudes have noticed that the starry sky appears to rotate about the Pole Star[+] once per day; the Chinese saw in this regularity the central role of their emperor, around whom all of life on Earth revolved. They incorporated this circular movement into instruments with rotating components, such as armillary spheres [B] and celestial globes [C].

Chinese astronomers emphasized the close relationship between heaven and Earth, particularly the belief that events in the heavens reflected those on Earth. Thus they saw new occurrences in the sky, such as novae[+] and comets, as indicators of important changes on Earth. They regarded the Milky Way[+] as the celestial counterpart of the Yellow River, linking them both to seasonal rainfall. As a centrally important heavenly object, the Milky Way sometimes appears flowing prominently through the middle of Chinese star charts [D].

Informed celestial interpretations required an intimate knowledge of the sky obtained in part by studying immense or detailed star charts [A]. Not surprisingly, belief in the parallels between heaven and Earth yielded an emphasis on the location of events in the sky rather than their brightness. Chinese celestial charts and the Japanese charts

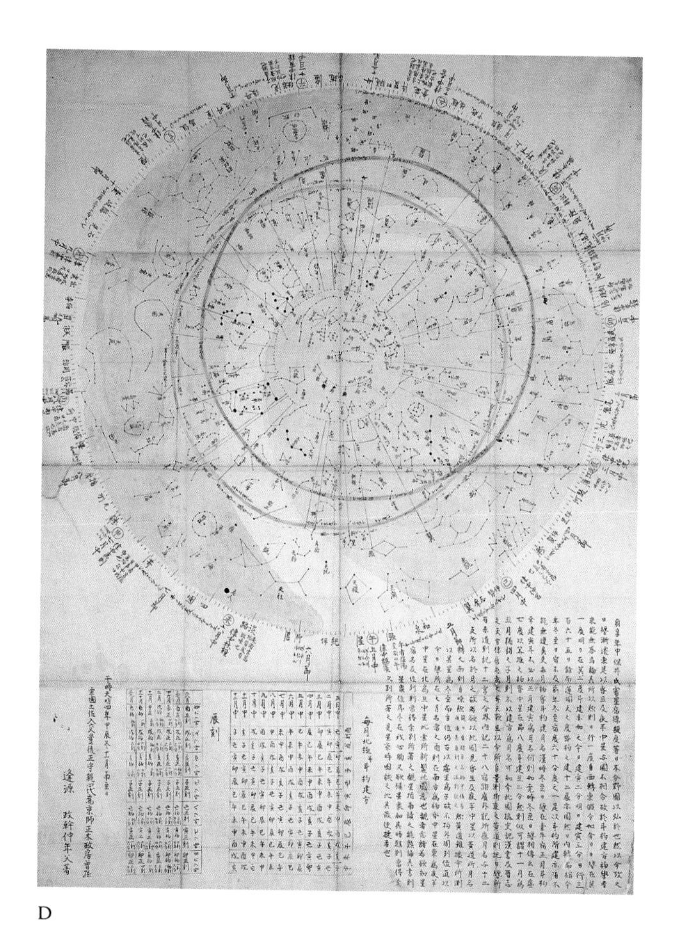

D

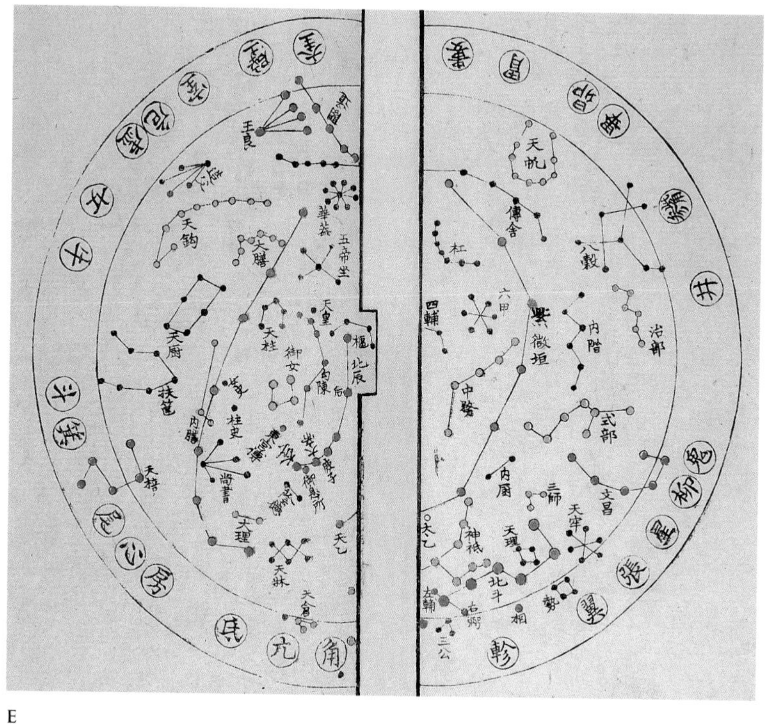

E

that derive from them can show nearly 300 groupings of stars, most of which bear little resemblance to those of other cultures and which do not match at all the magnitudes⁺ of stars as seen by the eye [F].

Chinese and Japanese star charts often depict these constellations, or asterisms,⁺ along with 28 lunar mansions [E], the borders of which all point toward the celestial pole.⁺ Each mansion originally marked the range of the Moon's daily motion. Because of precession,⁺ though, the mansions no longer equally divide the monthly lunar path.

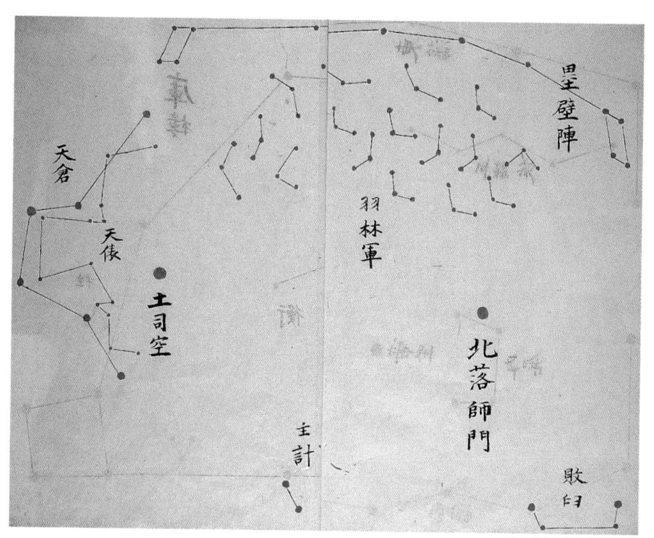

F

Observing Stars

What could be easier than observing the stars? People have always looked skyward, but locating the stars precisely, measuring their positions, is no simple task. For many measurements, early astronomers relied on an awkward "parallactic instrument" [B] that measured the distance of a star from the zenith, the point directly overhead. With spherical trigonometry, they could use these zenith distances to calculate a star's longitude and latitude (see appendix, p. 145 H). More convenient but often less accurate was the astrolabe (see pp. 64–67). The sight on the back of an astrolabe [A] or along a quadrant's edge allowed early astronomers to measure a star's altitude[+] above the horizon.[+] By the 1600s, astronomical quadrants sometimes carried telescopic sights, and even micrometers, for increased accuracy [C].

One direct approach to measuring a star's position was to observe it with an armillary sphere (see pp. 112–13) that had been fitted with sights. To achieve any accuracy, the armillary sphere had to be very large. Astronomers at the Peking observatory [D] used costly observing instruments like this, both before and after the arrival of Jesuit astronomers in the 1500s.

The obvious way to obtain more accuracy from a simple instrument such as a quadrant was to make it bigger, so that its user could read finer graduations on the scale. Skilled artisans made exceptionally large quadrants that were mounted

A

B

C

permanently (for stability) on north-south walls, and therefore were called mural quadrants. By using some kind of clock alongside a mural quadrant, an astronomer could measure the time and altitude of a star's culmination,[+] from which he could calculate its exact position. Such permanent installations often bore elaborate decorations: the mural quadrant of Tycho Brahe [E] enclosed a portrait of the astronomer and his household.

D

E

Unusual Events

Occasionally, spectacular and unusual events take place in the sky. During the day, solar eclipses turn the skies dark, and parhelia⁺ create multiple images of the Sun. At night, lunar eclipses cause the Moon to darken and turn colors, comets travel slowly across the sky, meteors appear as shooting stars, and novae⁺ suddenly materialize and then gradually fade. Believing in a connection between celestial occurrences and earthly happenings, people sought meaning in these unusual events. They tried to explain what they were, why they happened, and how they might affect daily life.

Eclipses

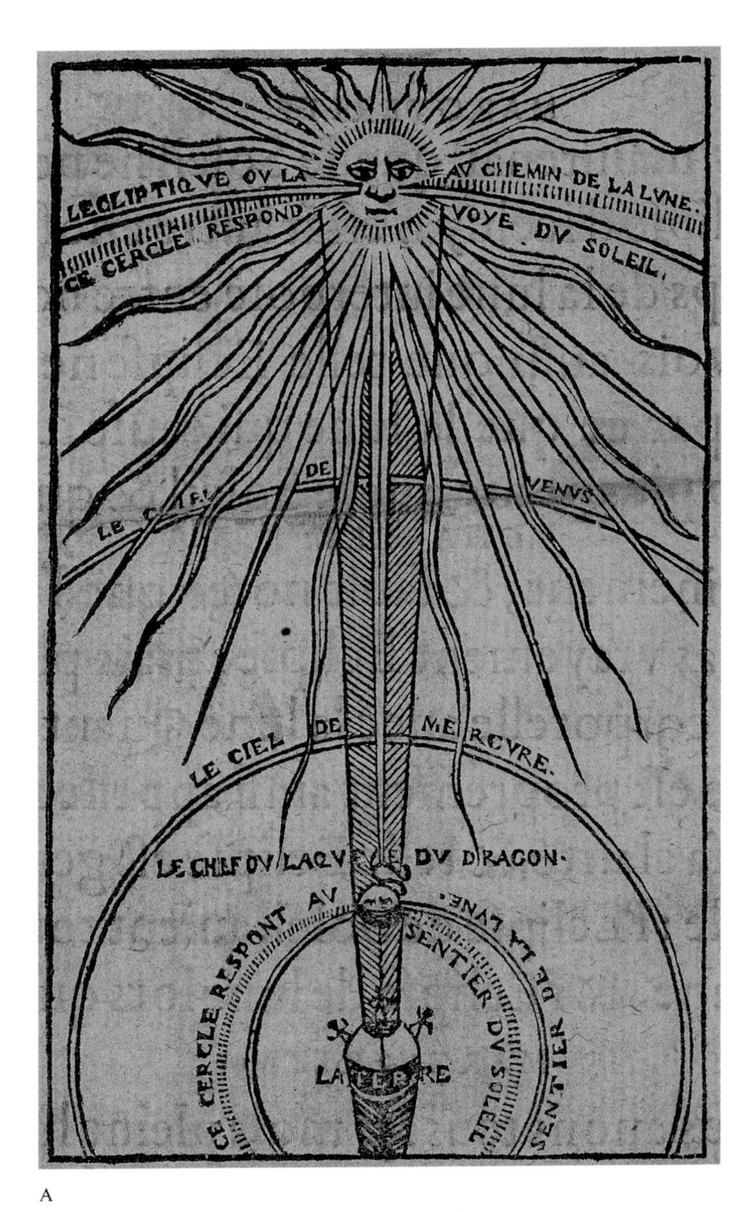

A

Prior to the 1900s, solar eclipses inspired fear and awe in most people as the sky darkened during daylight hours and the Sun disappeared behind a black disk. Lunar eclipses, though less spectacular, evoked similar mystery and amazement.

People have long observed and attempted to predict eclipses, desiring to understand why and how they happen. The earliest records of eclipses date more than 3000 years ago to ancient Babylonia and China. East Asian [B] and Islamic astronomers also recorded eclipses. The ancient Chinese thought that solar eclipses were caused by a dragon devouring the Sun. They attempted to scare off the dragon—by beating on drums and shooting off fireworks—to restore the sunlight.

In the ancient Greek world, Aristotle and Ptolemy understood the causes of eclipses to be tied to the geometry of the positions of the Sun, Moon, and Earth.

By the 1600s, many astronomical texts contained diagrams showing how solar and lunar eclipses occurred [A, E]. Craftsmen even created instruments to demonstrate the stages of an eclipse and to aid in calculating the timing of one [D].

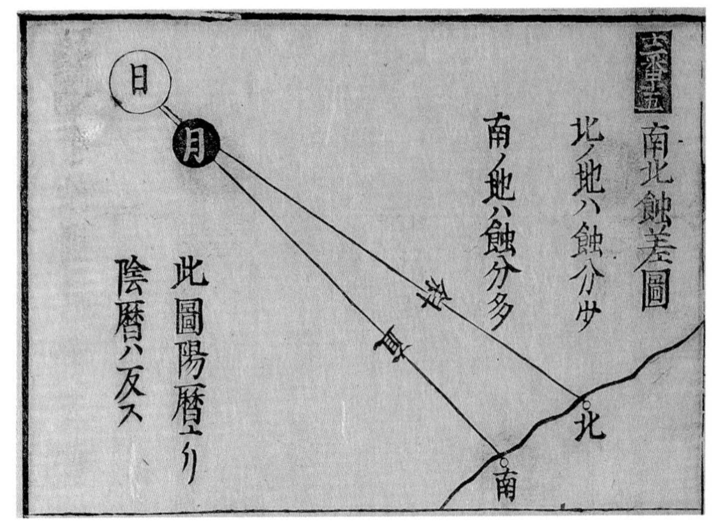

B

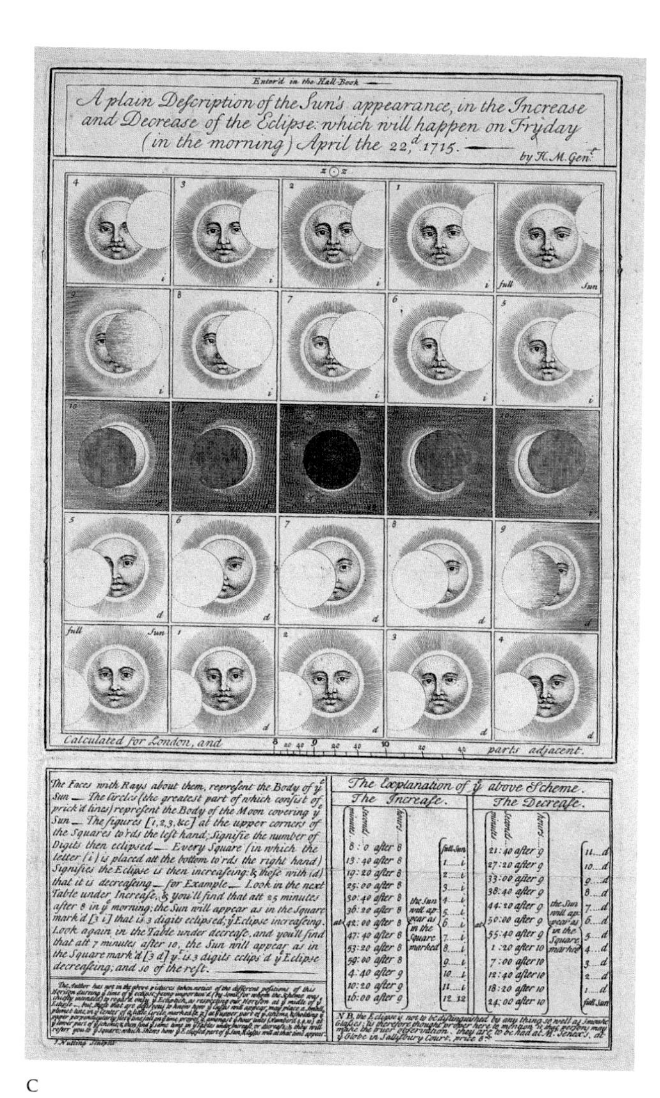

Enter'd in the Hall-Book

A plain Description of the Sun's appearance, in the Increase and Decrease of the Eclipse: which will happen on Fryday (in the morning) April the 22,d 1715. by H. M. Gent.

Calculated for London, and parts adjacent.

The Faces with Rays about them, represent the Body of ye Sun — The Circles (the greatest part of which consist of prick'd lines) represent the Body of the Moon covering ye Sun — The figures [1, 2, 3, &c] at the upper corners of the Squares to'rds the left hand; Signifie the number of Digits then eclipsed — Every Square (in which the letter (i) is placed att the bottom to'rds the right hand) Signifies the Eclipse is then increaseing: & those with (d) that it is decreaseing — for Example — Look in the next Table under Increase, & you'll find that att 25 minutes after 8 in ye morning; the Sun will appear as in the Square mark'd [3 i] that is 3 digits eclipsed; ye Eclipse increaseing. Look again in the Table under decrease, and you'll find that att 7 minutes after 10, the Sun will appear as in the Square mark'd [3 d] yt is 3 digits eclips'd ye Eclipse decreaseing; and so of the rest.

The Explanation of ye above Scheme.

The Increase. minutes : seconds, hours	the Sun will appear as in the Square marked	The Decrease. minutes : seconds, hours	the Sun will appear as in the Square marked
8 : 0 after 8	full Sun	21 : 40 after 9	11 ... d
13 : 40 after 8	1 ... i	27 : 20 after 9	10 ... d
19 : 20 after 8	2 ... i	33 : 00 after 9	9 ... d
25 : 00 after 8	3 ... i	38 : 40 after 9	8 ... d
30 : 40 after 8	4 ... i	44 : 20 after 9	7 ... d
36 : 20 after 8	5 ... i	50 : 00 after 9	6 ... d
42 : 00 after 8	6 ... i	55 : 40 after 9	5 ... d
47 : 40 after 8	7 ... i	1 : 20 after 10	4 ... d
53 : 20 after 8	8 ... i	7 : 00 after 10	3 ... d
59 : 00 after 8	9 ... i	12 : 40 after 10	2 ... d
4 : 40 after 9	10 ... i	18 : 20 after 10	1 ... d
10 : 20 after 9	11 ... i	24 : 00 after 10	full Sun
16 : 00 after 9	12 ... 12		

N.B. the Eclipse is not to be distinguished by any thing so well as Smoaky Glasses; tis therefore thought proper here to mention it that persons may make the truer observation. they are to be had at Mr. Senex's, at ye Globe in Sallsbury Court, price 6d.

J. Nutting Sculpsit

C

D

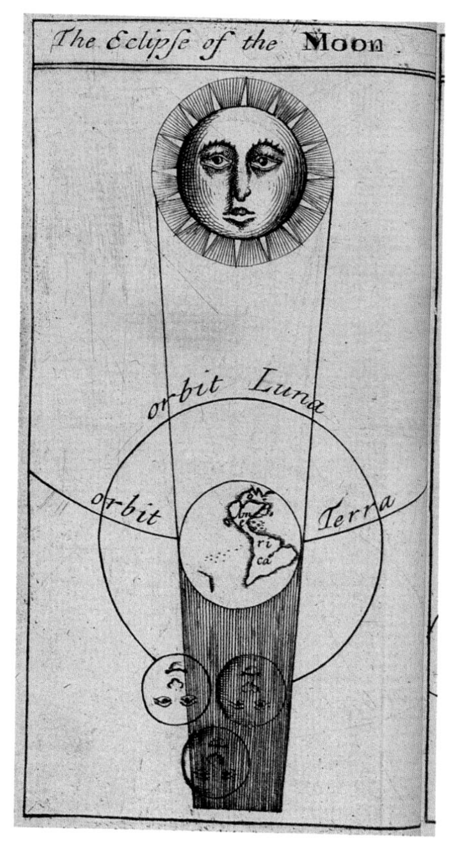

E

In addition to publishing books for astronomers, booksellers printed information sheets, or broadsides+ [C], that alerted and informed the public about eclipses. Some broadsides tried to explain the science behind eclipses in order to defuse public fear. Others took a more astrological approach to account for these remarkable phenomena.

Comets, Novae, Meteors

Unusual celestial occurrences other than eclipses terrified yet fascinated people of earlier eras. Our ancestors particularly feared comets, memorable because they appear so infrequently. A wide range of historical documents focused on these "blazing stars," especially in the 1600s, which witnessed many particularly bright comets.

Astronomers mapped cometary positions with respect to the fixed stars [A], drew detailed diagrams of comet shapes [B], and tried to figure out why they happened and where they came from. Broadsides alerted the public to the occurrence and possible implications of comets [C], and often focused on the astrological significance of a comet's presence in a particular part of the sky.

Novae⁺ are even rarer than comets. Early astronomers considered them to be emerging new stars [E] (*nova* means "new" in Latin), although they are actually exploding stars on their way to dying or variable stars whose brightness fluctuates.

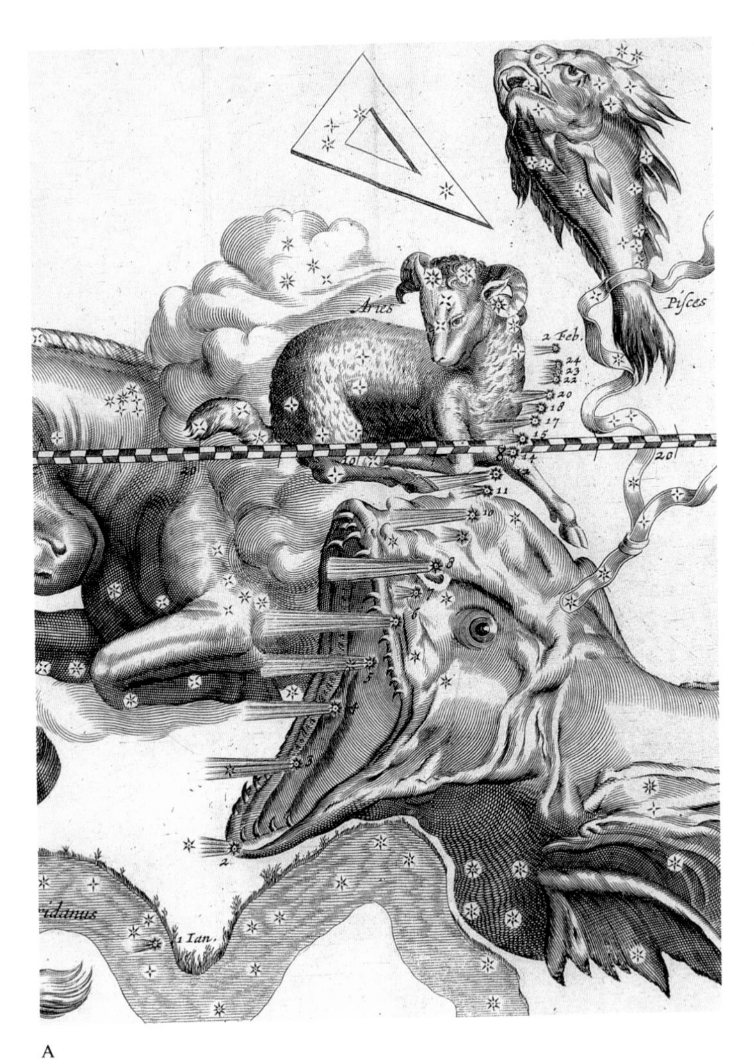

A

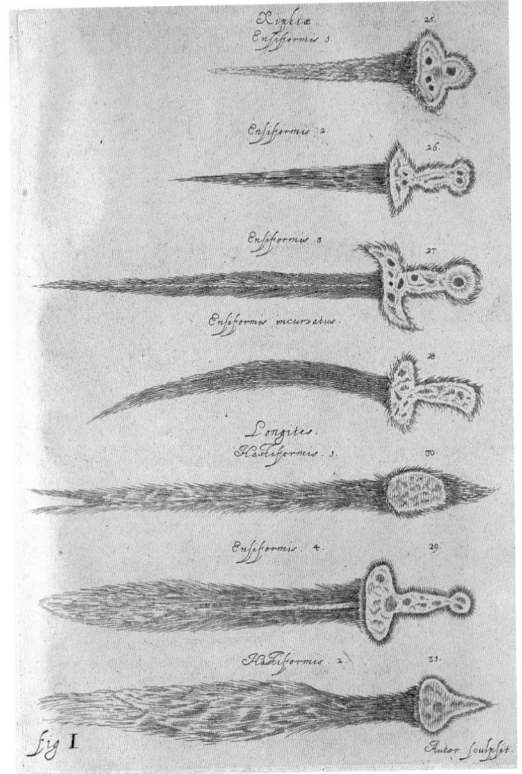

B

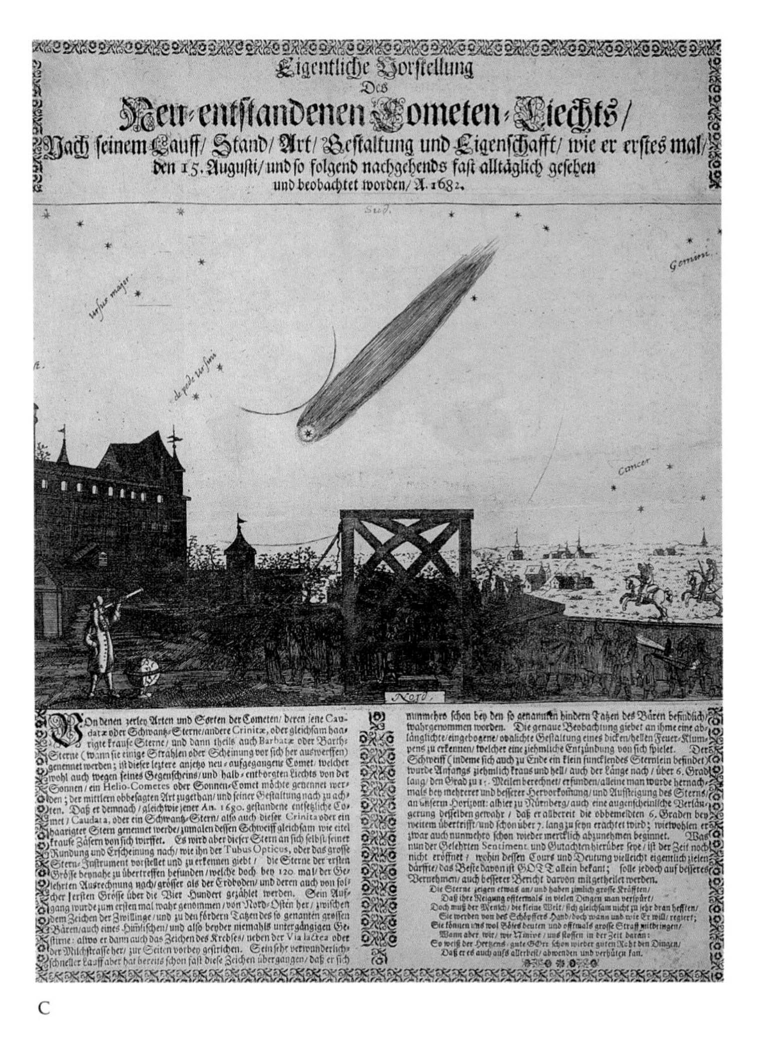

Eigentliche Vorstellung
Des
Neu-entstandenen Cometen-Liechts/
Nach seinem Lauff/ Stand/ Art/ Gestaltung und Eigenschafft/ wie er erstes mal/ den 15. Augusti/ und so folgend nachgehends fast alltäglich gesehen und beobachtet worden/ A. 1682.

Von denen zerley Arten und Sorten der Cometen/ deren jene Caudatæ oder Schwantz-Sterne/ andere Crinitæ, oder gleichsam haarigte krause Sterne/ und dann theils auch Barbatæ oder Barth-Sterne (wann sie einige Strahlen oder Scheinung vor sich her auswerffen) genennet werden; ist dieser leztere anjetzo neu-aufgegangene Comet/ welcher wohl auch wegen seines Gegenscheins/ und halb-entborgten Liechts von der Sonnen/ ein Helio-Cometes oder Sonnen-Comet möchte genennet werden; der mittlern obbesagten Art zugethan/ und seiner Gestaltung nach zu achten. Daß er demnach/ gleichwie jener An. 1680. gestandene entsetzliche Comet/ Caudata, oder ein Schwantz-Stern/ also auch dieser Crinita oder ein haarigter Stern genennet werde/ zumalen dessen Schweiff gleichsam wie eitel krause Zäsern von sich wirffet. Es wird aber dieser Stern an sich selbst/ seiner Rundung und Erscheinung nach/ wie ihn der Tubus Opticus, oder das grosse Stern-Instrument vorstellet und zu erkennen giebt/ die Sterne der ersten Grösse beynahe zu übertreffen befunden/ welche doch bey 120. mal/ der Gelehrten Ausrechnung nach/ grösser als der Erdboden/ und deren auch von solcher ersten Grösse über die Vier Hundert gezählet werden. Sein Aufgang wurde zum ersten mal wahr genommen/ von Nord-Osten her/ zwischen dem Zeichen der Zwillinge/ und zu den fördern Tatzen des so genanten grossen Bären/ auch eines Hünlischen/ und also beyder niemahls untergängigen Gestirne: allwo er dann auch das Zeichen des Krebses/ neben der Via lactea oder der Milchstrasse her/ zur Seiten vorbey gestrichen. Sein sehr verwunderlich-schneller Lauff aber hat bereits schon fast diese Zeichen übergangen/ daß er sich nunmehro schon bey den so genanten hindern Tatzen des Bären befindlich/ wahrgenommen worden. Die genaue Beobachtung giebet an ihme eine ablänglichte/ eingebogene/ ovalichte Gestaltung eines dicken/ hellen Feuer-Klumpens zu erkennen/ welcher eine ziehmliche Entzündung von sich spielet. Der Schweiff (indeme sich auch zu Ende ein klein funcklendes Sternlein befindet) wurde Anfangs ziehmlich kraus und hell/ auch der Länge nach/ über 6. Grad lang/ den Grad zu 1 :. Meilen berechnet/ erfunden/ alleine man wurde hernachmals bey mehrerer und besserer Hervorkommung/ und Auffsteigung des Sterns/ an unserm Horizont: alhier zu Nürnberg/ auch eine augenscheinliche Verlängerung desselben gewahr/ daß er allbereit die obbemeldten 6. Graden bey weitem übertrifft/ und schon über 7. lang zu seyn erachtet wird; wiewohlen er zwar auch nunmehro schon wieder merklich abzunehmen beginnet. Was nun der Gelehrten Sentiment und Gutachten hierüber seye/ ist der Zeit noch nicht eröffnet/ wohin dessen Cours und Deutung vielleicht eigentlich zielen dörffte/ das Beste davon ist GOTT allein bekant; solle jedoch auf besseres Vernehmen/ auch besserer Bericht davon mitgetheilet werden.

Die Sterne zeigen etwas an/ und haben zimlich grosse Kräfften/
Daß ihre Neigung offtermals in vielen Dingen man verspürt/
Doch muß der Mensch/ die kleine Welt/ sich gleichsam nicht zu sehr dran hefften/
Sie werden von des Schöpffers Hand/ doch wann und wie Er will/ regiert;
Sie können uns wol Böses deuten und offtmals grosse Straff mitbringen/
Wann aber wir/ wie Ninive/ uns stossen in der Zeit daran:
So weiß der Hertzens-gute GOtt schon wieder guten Raht den Dingen/
Daß er es auch aufs allerbest/ abwenden und verhüten kan.

C

Meteors occur as single events and in annual showers [D], the biggest of which, the Perseids and the Leonids, happen in August and November respectively. Though the common name for a meteor, shooting star, hints at its origins, only by the mid-1800s did astronomers begin to understand what produces these streaks of light—gravity draws small objects from space into the Earth's atmosphere, where friction burns them up as they are pulled to the surface of the Earth. The heat generated from this process causes the gases in the air around them to glow.

D

E

Navigating by the Stars

In the 1400s, Portuguese and other sailors began the purposeful discovery and exploration of distant lands. With no landmarks available, the seafarer found his location by taking note of the positions of the stars. Because the night sky changes moment by moment and from night to night, only an intimate knowledge of the stars and their movements proves sufficient to find one's whereabouts on the high seas. To this end, an improved class of navigational instruments emerged during the Renaissance, in support of the new and wide-ranging commercial, political, and colonial interests.

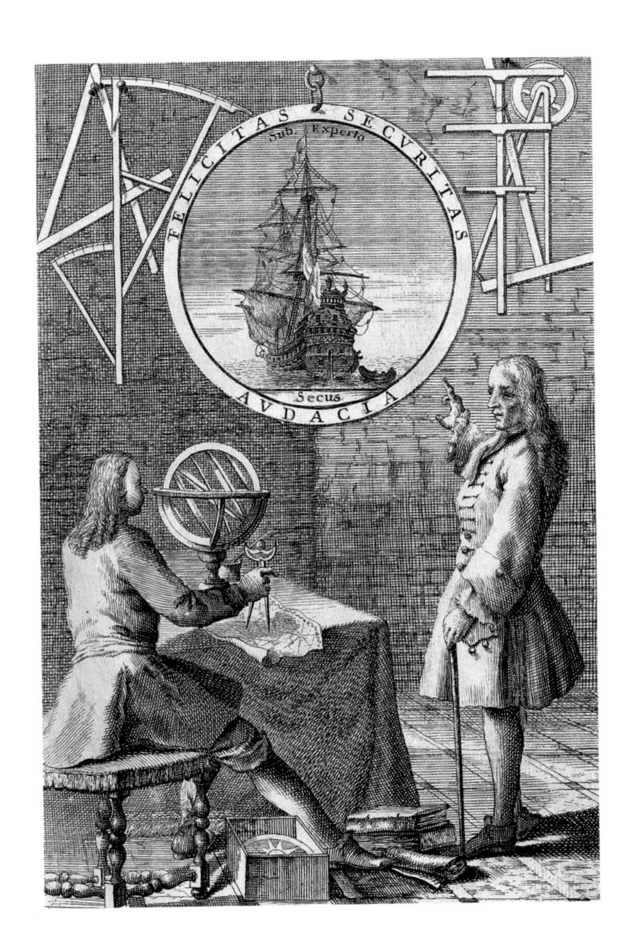

Compasses and Mariner's Astrolabes

Explorers returning from China introduced magnetized needles to Europeans around 1200. Attaching such a needle to the bottom of a freely rotating circular card created the mariner's compass, which evolved into ever more elaborate forms [A]. Its compass card divides the horizon into 32 points, or rhumbs [B]. The earliest navigation maps, called portolan charts, included a similar network of rhumb lines. By choosing the line most closely connecting two points on the chart, and then sailing along the same rhumb on the compass card, a navigator could maintain a steady course even when overcast skies obscured heavenly bodies. Portolan charts indicated bays, rocks, and other noteworthy landmarks that could help locate the ship's position along a coast.

A

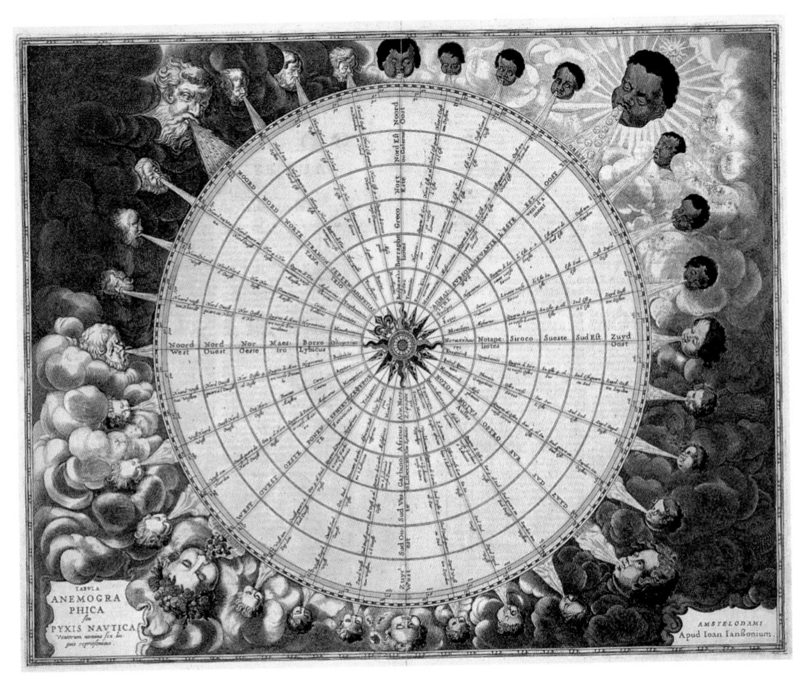

B

C

D

But seafarers negotiating vast expanses of open sea and ocean discovered the inadequacy of portolan charts. Beginning in 1438 at Sagres, Portugal, the fledgling science of positional navigation yielded different navigational charts and new instruments to use with them. With this system, sailors located their position at sea according to latitude,[+] the distance north or south from the equator, and longitude,[+] the distance east or west from an agreed point, such as the meridian[+] running through Greenwich, England.

Determining one's latitude is quite easy. The simplest technique involves finding the height, or altitude,[+] of the Pole Star[+] above the horizon;[+] when measured in degrees, that altitude nearly equals the local latitude. Although quadrants had been used to find such information since Roman times and worked well on land, they could be difficult to employ on board a heaving ship. This problem, as well as exploration south of the equator, where the Pole Star is not visible, led to the development of navigational instruments and techniques for converting the noontime altitude of the Sun to local latitude. One such instrument was the mariner's astrolabe [C], of which fewer than 100 are known, nearly all of them recovered from shipwrecks.

Devising a method for finding local longitude vexed astronomers and navigators until the 1700s (see p. 93). As a result, maps until then often show marked distortions in east-west distances [D]. More important, sailors could easily get lost, placing them in danger of falling ill, running out of supplies, or sailing into unnoticed hazards.

The Cross-Staff and the Back-Staff

The wooden cross-staff [A], also known as the Jacob's staff or the forestaff, originated in the 1300s. This device, a 30-inch, four-sided calibrated staff along which one or more vanes move, measures angles on Earth or in the sky. By placing the staff near one's eye, and then sliding an adjustable vane until its ends align with the top and bottom of a building, for example, one can determine the structure's height mathematically. Later users employed the cross-staff to measure the angle between two celestial features, such as the Moon and a star, or the elevation of an object above the horizon⁺ [B].

By around 1515, Portuguese sailors used the cross-staff at sea to find the altitude⁺ of the Pole Star.⁺ The seagoing instrument might include three or four vanes of varying sizes. Due to

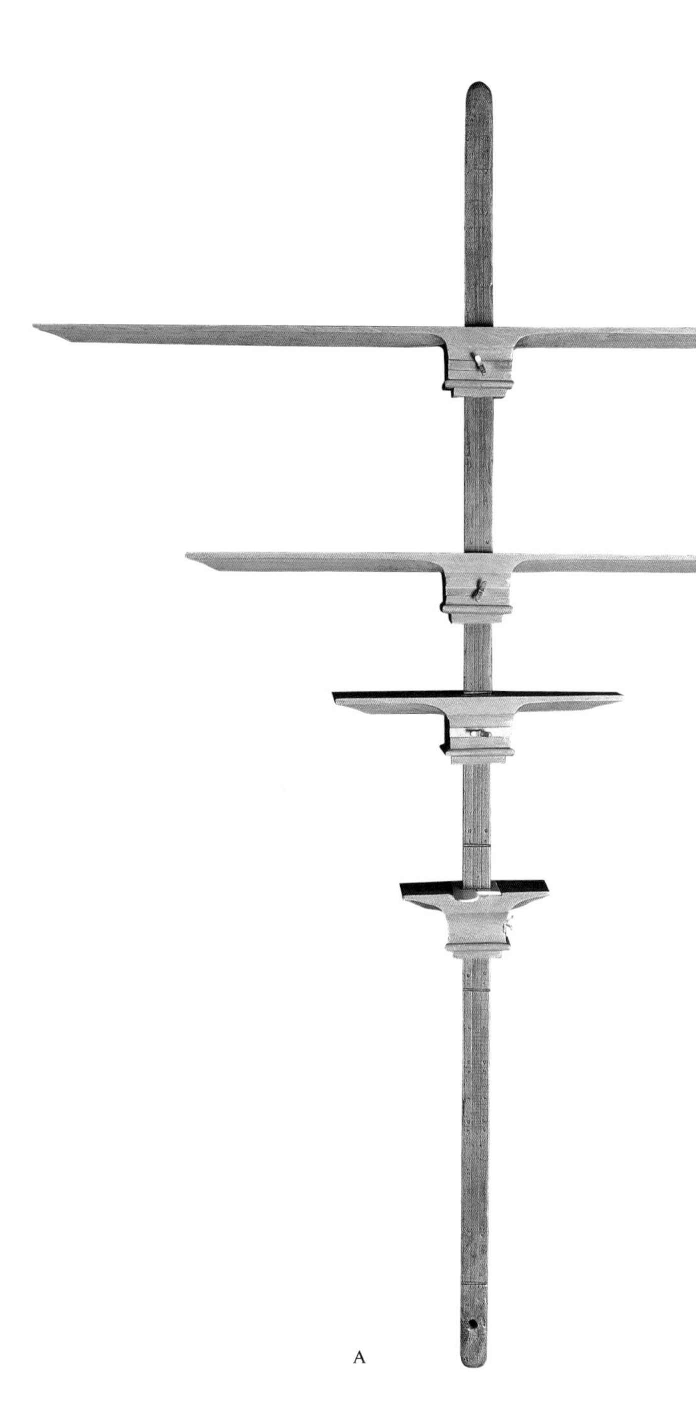

A

B

C

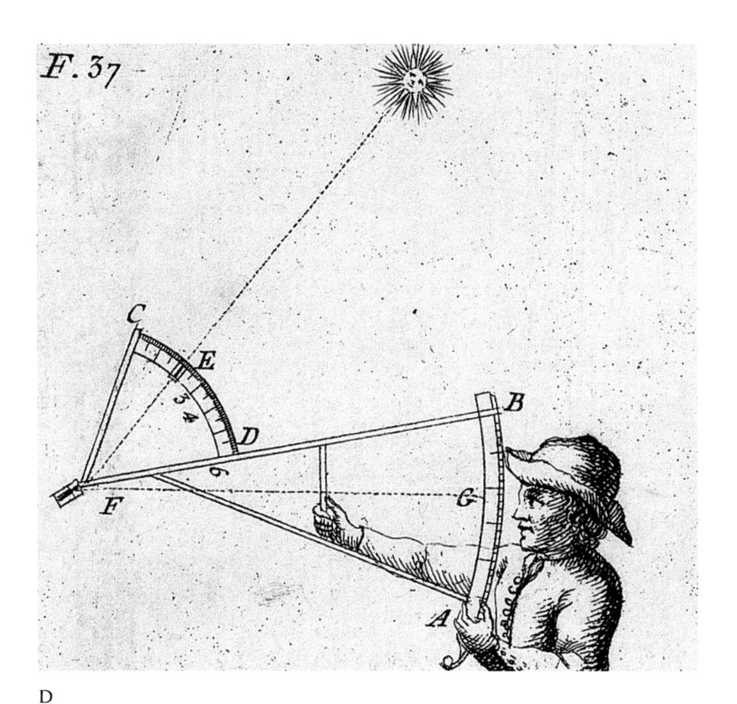

D

the obvious danger of looking directly into the Sun when using a cross-staff [B], sailors preferred the mariner's astrolabe [C] or nautical quadrant for sighting the Sun.

In 1595, the English captain John Davis introduced the back-staff [E], also known as the Davis or English quadrant. It avoided harm to the eyes by keeping the observer's back to the Sun [D]. It also eliminated the tiresome technique of sighting quickly to both ends of the cross-staff's vane. Adjustment of the upper vane of the back-staff placed the Sun's shadow at the far end of the staff, with the second vane nearest the eye aligning with the horizon. Adding the angles indicated on each sector gave the Sun's angular elevation.

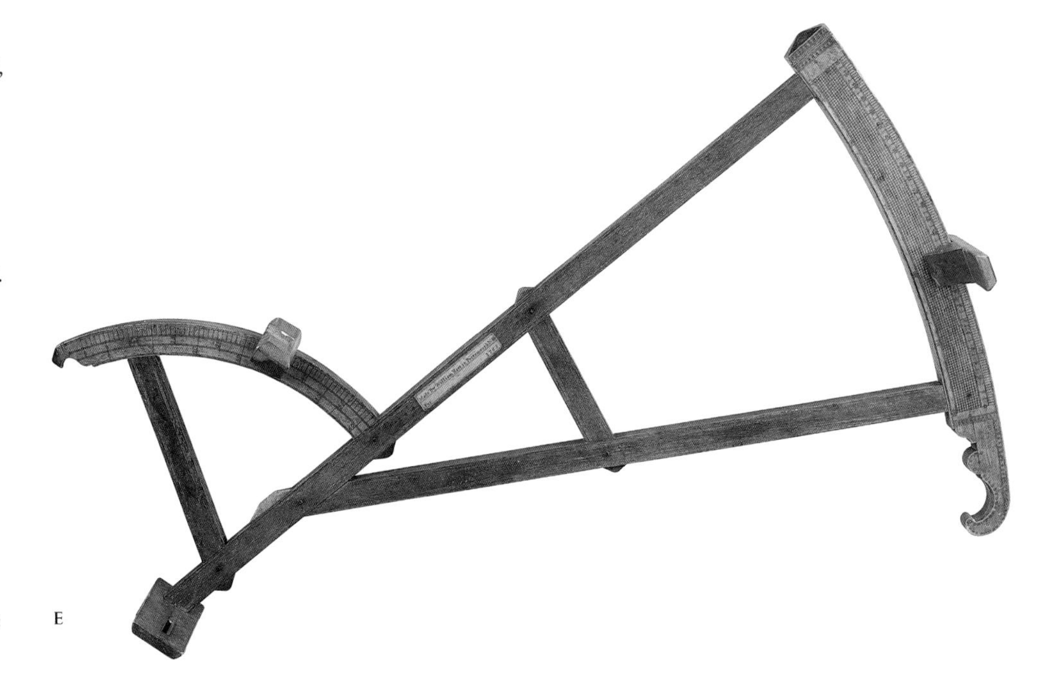

E

Quadrants, Sextants, Octants

People sometimes erroneously use the word "quadrant" to refer to a quadrant, sextant, or octant because each instrument can measure a quarter circle (or more, in the case of the sextant). But the proper name of each of these devices refers to the size of its arc, marked with a scale of degrees. Thus a quadrant consists of an arc forming one-quarter of a circle, a sextant one-sixth of a circle, and an octant one-eighth of a circle.

To use a quadrant, one views a distant object along one of the straight edges, which is fitted with a pair of sights. The arcs of early Portuguese nautical quadrants marked locations corresponding to various elevations of the Pole Star.[+] On the curved edge of later quadrants, a plumb bob indicated the angle of an object's elevation above the horizon[+] [A].

In 1731, John Hadley presented a new and much more accurate instrument [B], an octant known confusingly as Hadley's quadrant. Additional improvements, such as a more rigid frame and a more precisely calibrated scale, yielded even

A

B

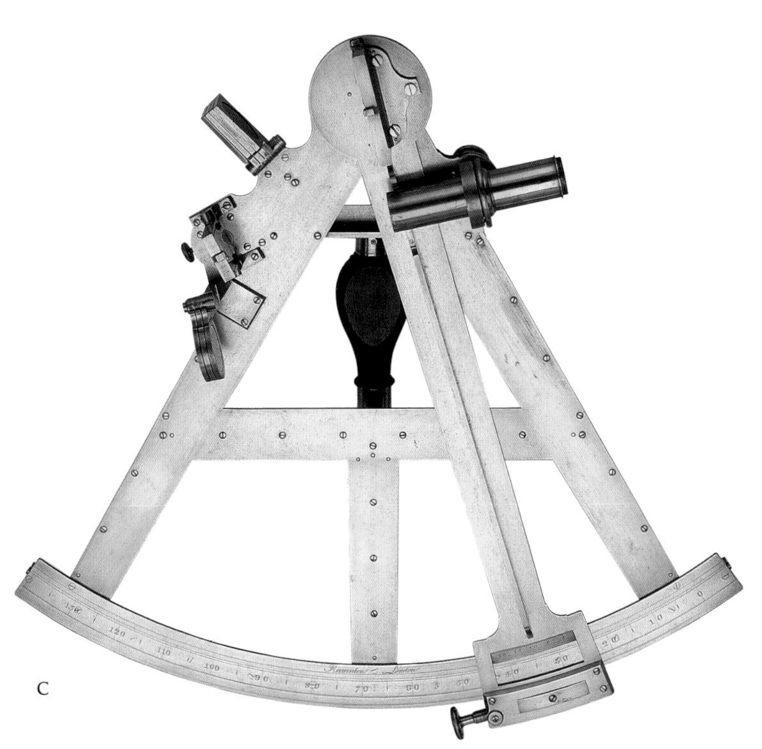

C

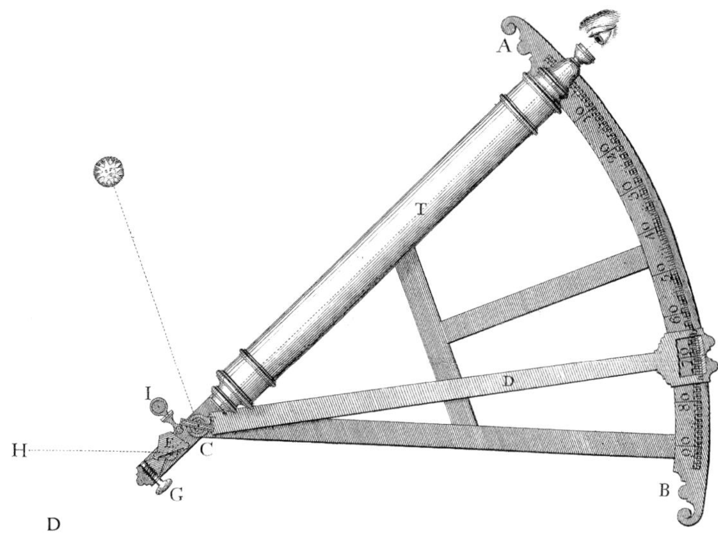

D

greater accuracy. By 1770, the desire to measure larger celestial angles in order to use the lunar distance solution to the longitude problem (see p. 93) led to the expansion of the octant's arc to one-sixth of a circle, producing the sextant [C].

The key improvement of the octant and sextant over earlier devices derived from their use of mirrors [D]. The first mirror is located on a radial arm that moves along the arc of the instrument. The navigator adjusts this arm until the mirror reflects the image of the Sun (or Moon) toward the silvered half of a second mirror. Looking through a shaded pinhole sight, the navigator's eye simultaneously views the doubly reflected Sun and the horizon appearing through the clear half of the second mirror. The instrument's arc, marked in degrees, then indicates the angle of elevation of the Sun above the horizon. A similar technique provided the angular separation between the Moon and a reference star for the determination of longitude.[+] The repeating circle [E] similarly uses mirrors to take numerous sightings, adding the angle found each time. Dividing this total angle by the number of sightings results in a mean or average value assumed to be far more accurate than any single measurement.

E

Telescopic Investigation

The telescope opened up the skies to deeper scrutiny. The universe it revealed is unimaginably huge yet works by the same rules that operate on Earth. Early telescopes, by comparison with the giant instruments astronomers use today, were skinny, but for good reason. Unless they are quite thin, simple lenses distort images. Because thin lenses require a greater distance to bring an image into focus, some early telescopes are so long that they actually omit the tube. Only the thin lens is left, high in the sky, with an observer on the ground trying forlornly to line up an eyepiece with it. Wonderful things were there to see, nonetheless, and wonderful things the early observers saw.

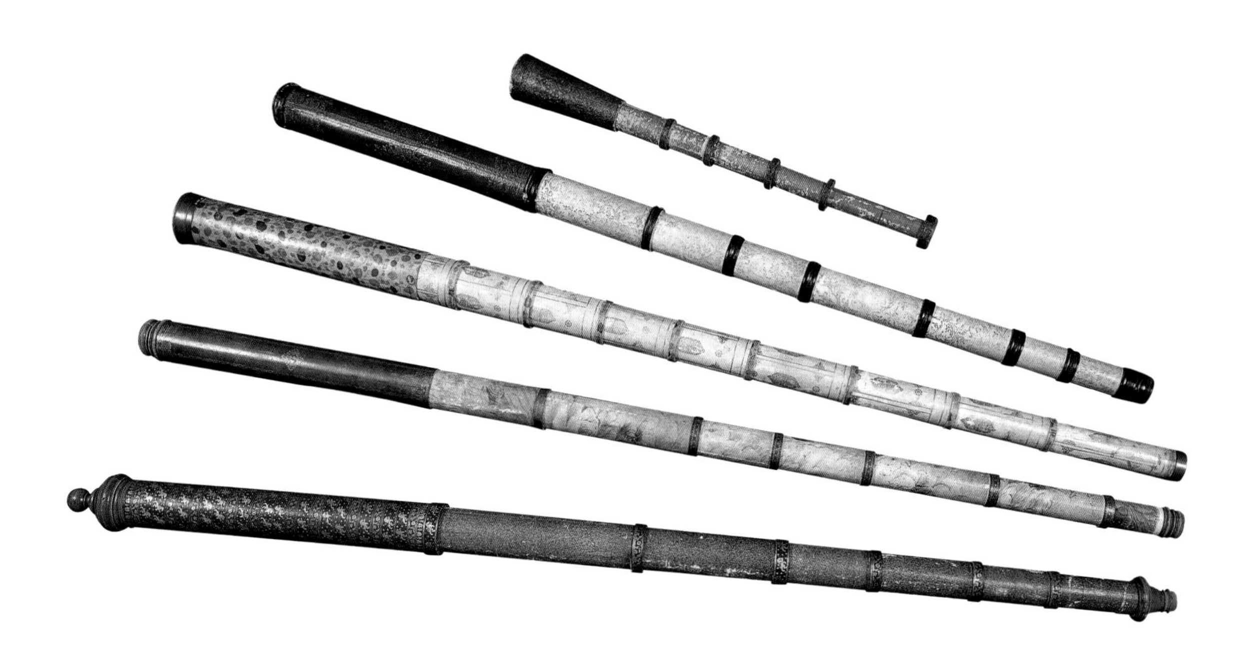

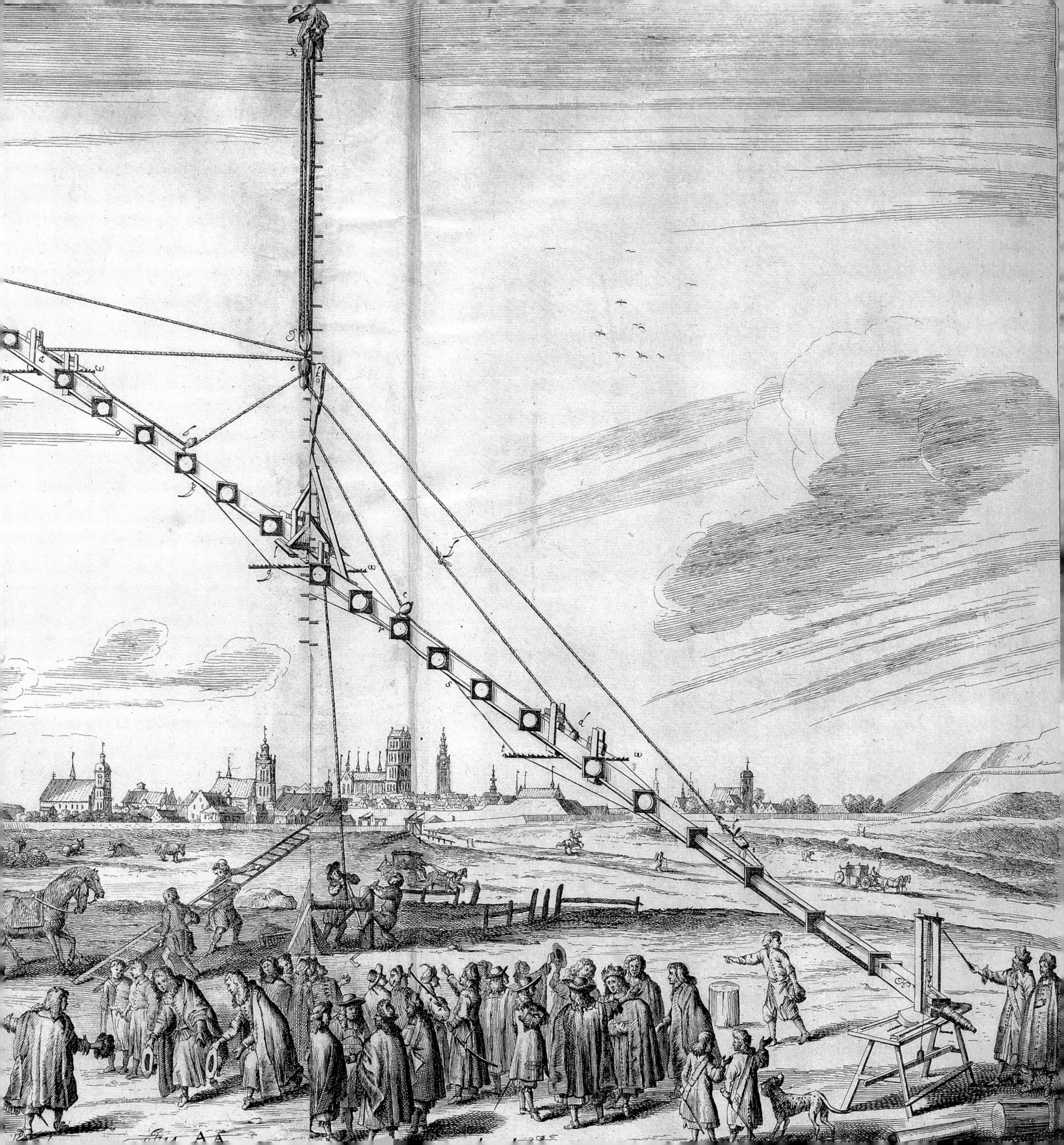

Surface of the Moon and New Moons

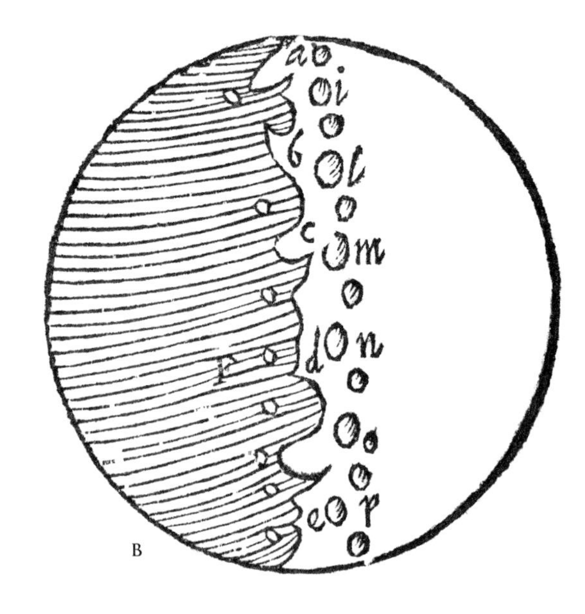

B

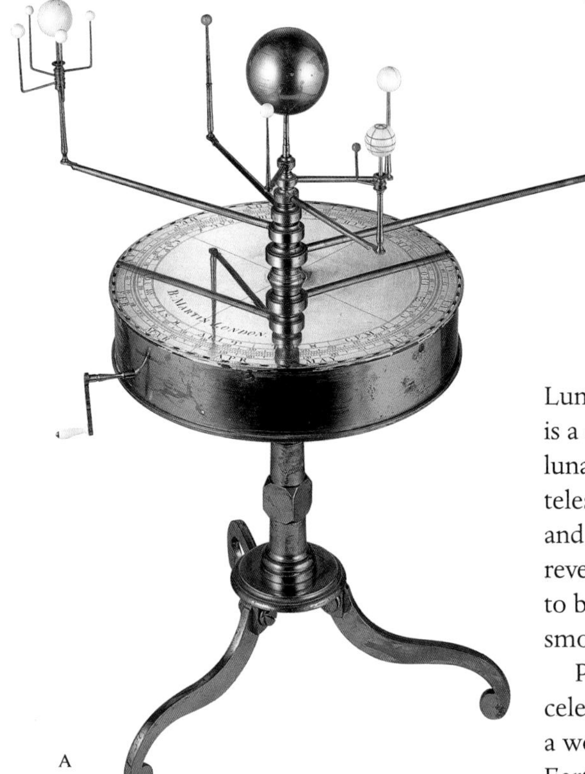

A

Lunar landscape: today the very phrase is a cliché. People in the 1600s found the lunar landscape a startling novelty. The telescopes of Galileo, Riccioli, Hevelius, and other observers for the first time revealed the terrain of the Moon [B, C] to be mountainous and cratered, not smooth.

People had always thought the celestial world unchanging and perfect, a world fundamentally different from Earth. The first drawings of the Moon profoundly unsettled them. Astronomers published the drawings over and over [E], driving home the first revolutionary message unveiled by the telescope. The Moon was not the perfect celestial body that philosophers described. In fact, it looked much like the Earth.

Meanwhile, astronomers realized that the Earth's moon was not the only one in the heavens. Sequential arrangement of Galileo's telescopic drawings from January 1610 reveals this [D]. Jupiter appeared in the crude telescope as no more than a bright point of light. Fainter points of light also were visible, all in a row, as many as four of them following Jupiter through the sky. From night to night, the points of light danced. They swapped places, moving now east, now west of the central planet, sometimes hiding altogether.

After many nights, the message became clear: Jupiter has moons that orbit around it. We could no longer consider the Earth to be the center of all motion in the heavens. And if Earth is not the center of all heavenly motion,

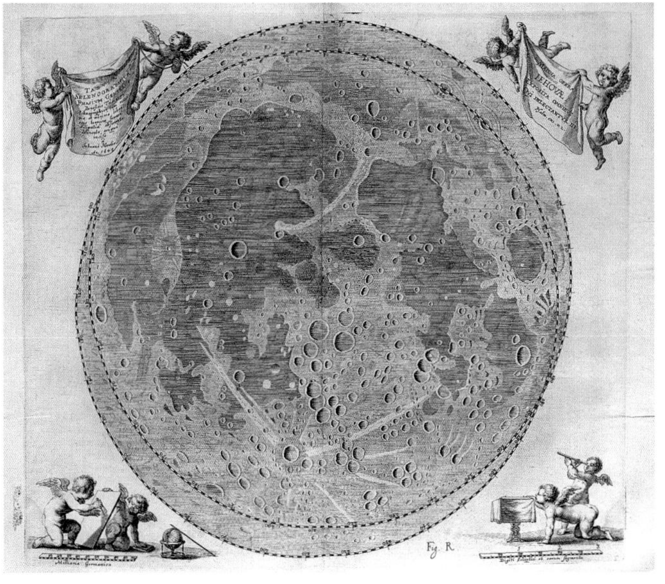

C

MOEDICEORVM PLANETARVM
ad inuicem, et ad IOVEM Constitutiones, futuræ in Mensibus Martio
et Aprile An: M DCXIII. à GALILEO G.L. earundem
Stellarũ, nec non Periodicorum ipsarum motuum
Repertore primo, Calculis collectæ ad
Meridianum Florentiæ

Martij
Die 1. Hor. 3 ab Occasu.
Hor. 4.
Hor. 5.
Die 2 H. 3
Die 3 H. 3
Die 4 H. 3.
Die 5 H. 2.
H. 3 Pars versus Ortum Pars versus occ.

D

how could we know it was at the center of the heavens? This ultimate message of Galileo's sequence of telescopic drawings forever changed our view of the cosmos. In 1655, Christiaan Huygens detected a moon around Saturn, and eventually Saturn's moons showed up in models of the solar system [A].

Extraterrestrial moons did not prove that our home the Earth flies through space like a planet, but they broadened our concept of a planet. It became easier and easier to imagine Planet Earth.

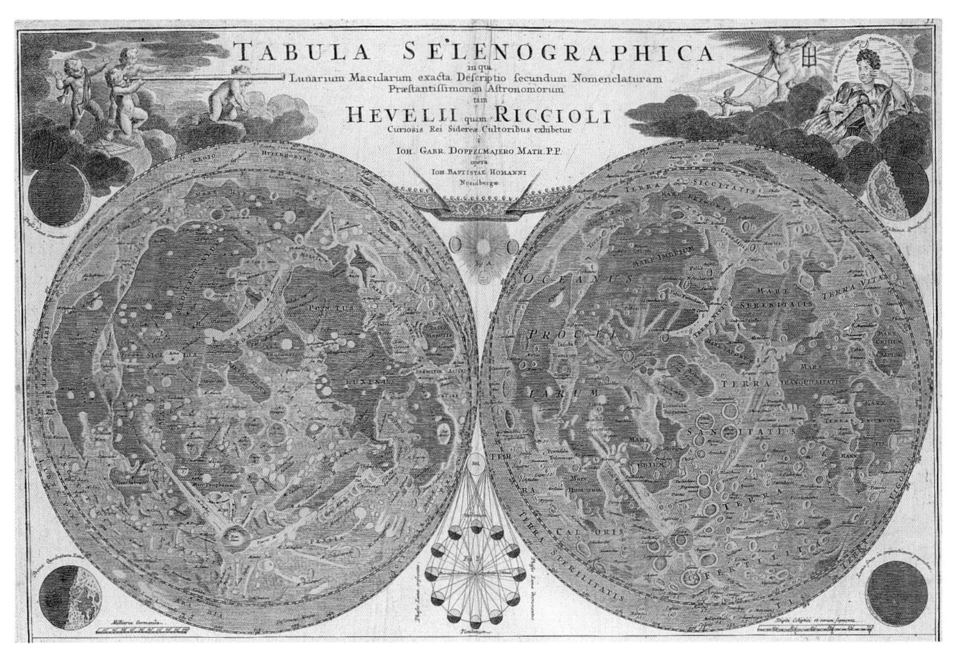

E

Distant Worlds Discovered and Unveiled

In 1781, the telescope revealed another surprise. Using a reflecting telescope of his own design [A], musician William Herschel serendipitously noticed an unusual object. The only thing he thought it could be was a comet. Astronomers showed that it was, in fact, a new planet, the first one discovered in recorded history, orbiting far beyond Saturn. Herschel referred to it as Georgium Sidus, Latin for "George's Star," in honor of King George III of England (see p. 135 D); others called it Herschel after its discoverer. It took astronomers about 50 years to agree to name it after Jupiter's mythological father, Uranus.

On New Year's Day 1801, Italian astronomer Giuseppe Piazzi discovered a different sort of unusual object. This one orbited between Mars and Jupiter. Other astronomers soon found more of these small planetlike bodies, which Herschel dubbed "asteroids," the name by which we call them today. For decades, asteroids appeared in lists of planets and on diagrams of the solar system [B]. The telescope continued to

A

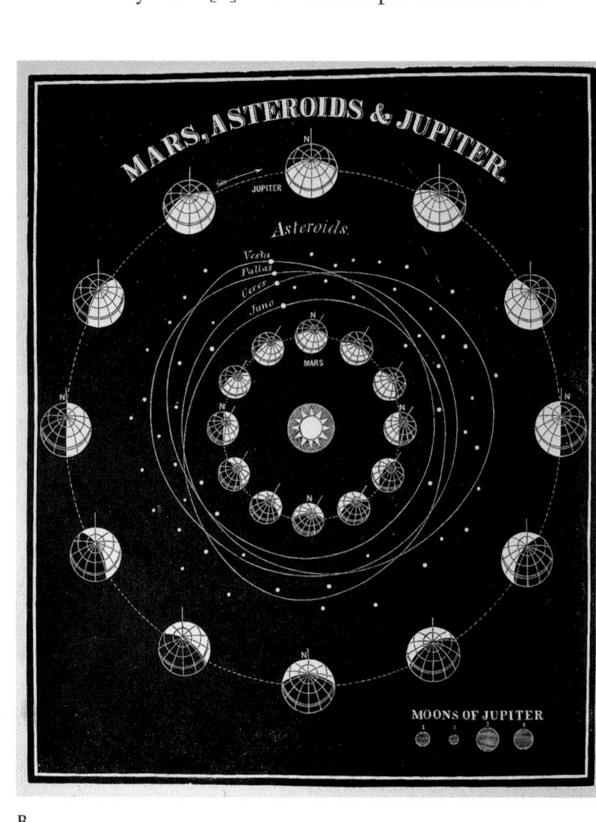

B

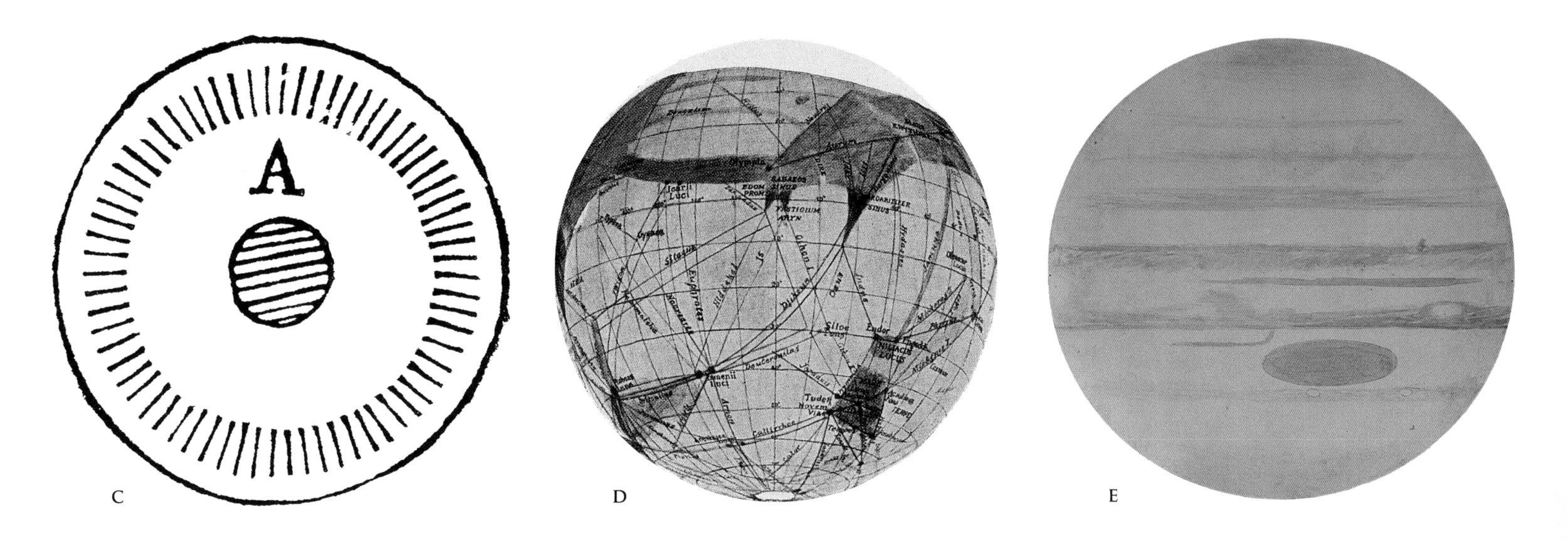

C D E

reveal more and more worlds, including another very distant planet, Neptune, discovered in 1846 after mathematical astronomers predicted its existence and location.

Early telescopes also hinted at planetary details, such as the surprising and changing appearance of appendages on Saturn. In a series of sketches in the 1650s, Christiaan Huygens showed how a flattened ring surrounding Saturn, and inclined to its orbital plane, could explain these strange phenomena [F]. The earliest drawings made of the Martian surface, dating from the 1630s [C], reflected imperfections in the observer's telescope lenses rather than features on Mars. Later sketches, with their crude markings of the Martian surface, suggested that it and other planets had features like those on Earth. By the late 1800s, astronomers sketched the Martian surface with increasing detail [D].

At that time, telescopic views of Jupiter still revealed little more than a few bands and its great Red Spot [E]. But this was enough to suggest the likelihood of immense, violent storms there, and to challenge astronomers and others to continue studying this cold, distant, and inhospitable world.

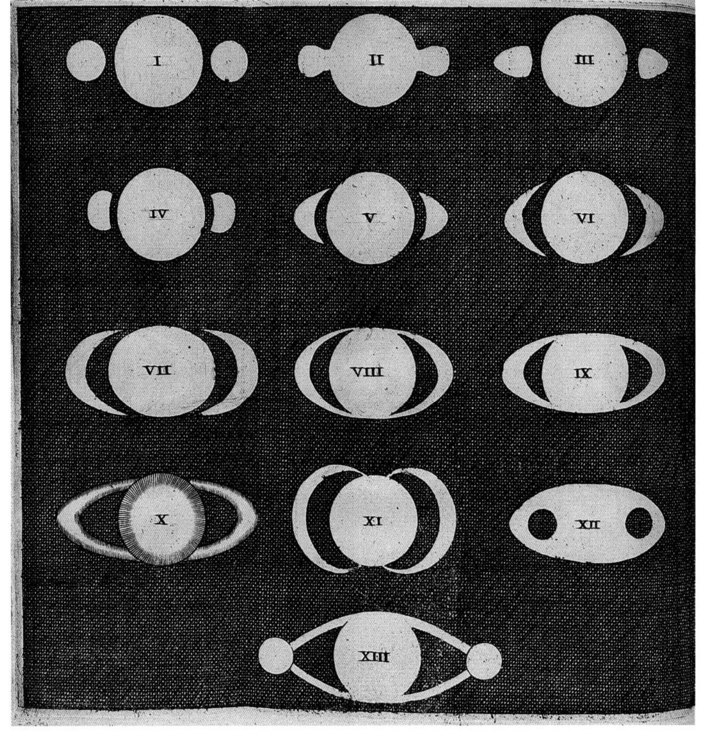

F

Telescopes and Stars

After the development of telescopes, astronomers could observe many stars invisible to the naked eye. In 1729, John Flamsteed published the first comprehensive telescopic star catalogue+ and a companion celestial atlas+ [A] that carefully plotted these new stars.

Stars now filled areas of the sky that previously had seemed empty. Often celebrating recent inventions, astronomers added new constellations+ [B] that departed from earlier mythological identities and sometimes consisted solely of faint stars or stars invisible to the naked eye. Cartographers also mapped many more stars within existing constellations.

Artistic celestial mapping culminated in Johann Elert Bode's *Uranographia*—the last great pictorial celestial atlas, published in 1801 [C]. As astronomers

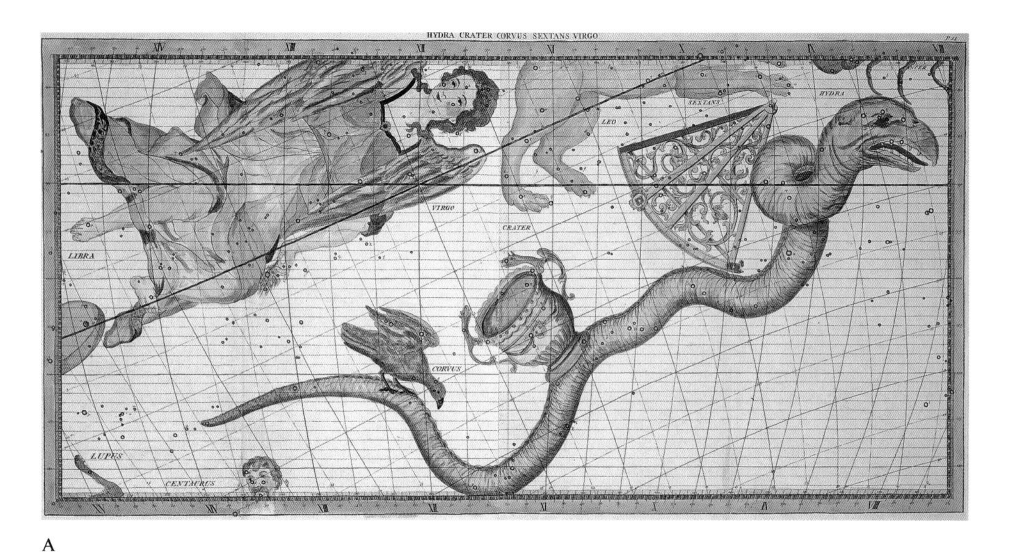

A

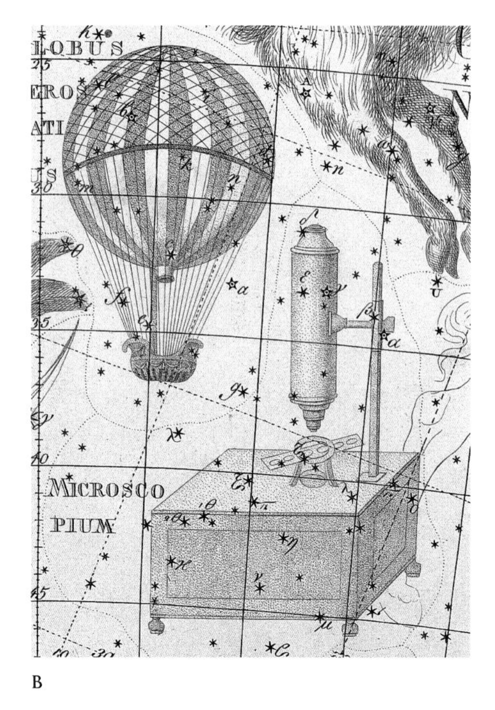

B

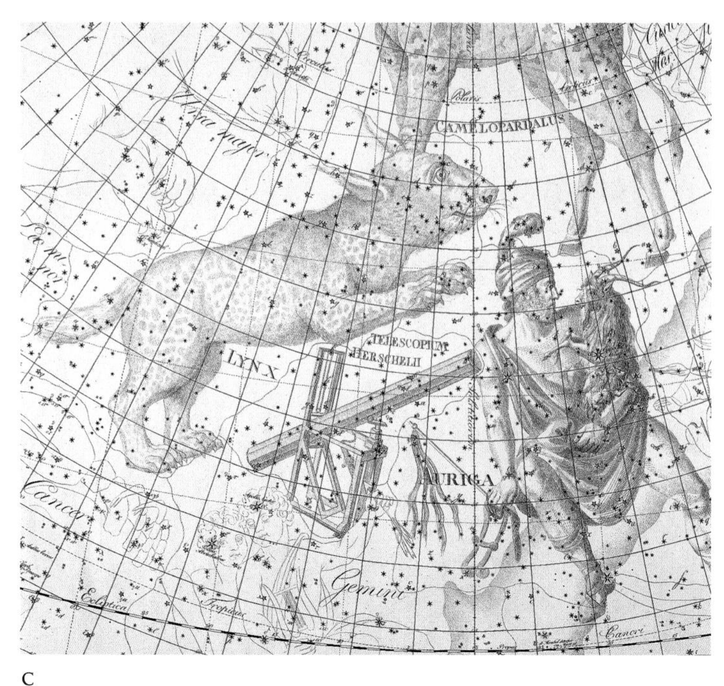

C

developed more powerful telescopes and catalogued increasing numbers of stars, pictorial celestial atlases became impractical. The outlines of the constellation figures would have obscured the tiny dots plotted to represent the stars.

During the 1800s, two categories of star charts evolved. Those for astronomers featured the stars as austere points on a map [F]. Sometimes boundary lines divided the sky into regions defined by their old constellation names. Charts for the general public featured stars visible only to the naked eye [G], sometimes delineating faint outlines of the constellations. At first, such outlines resembled earlier star charts [D], but later they took on more geometric shapes that focused on easily recognizable asterisms⁺ [E].

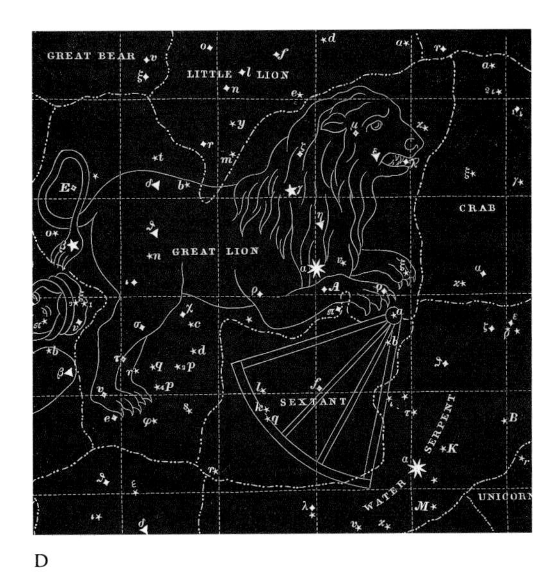

D

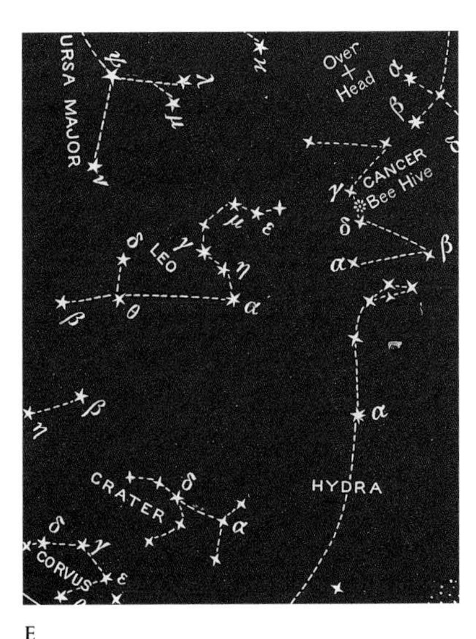

E

CARTE PHOTOGRAPHIQUE DU CIEL

Zone + 7° N° 67

Observatoire de Toulouse

F

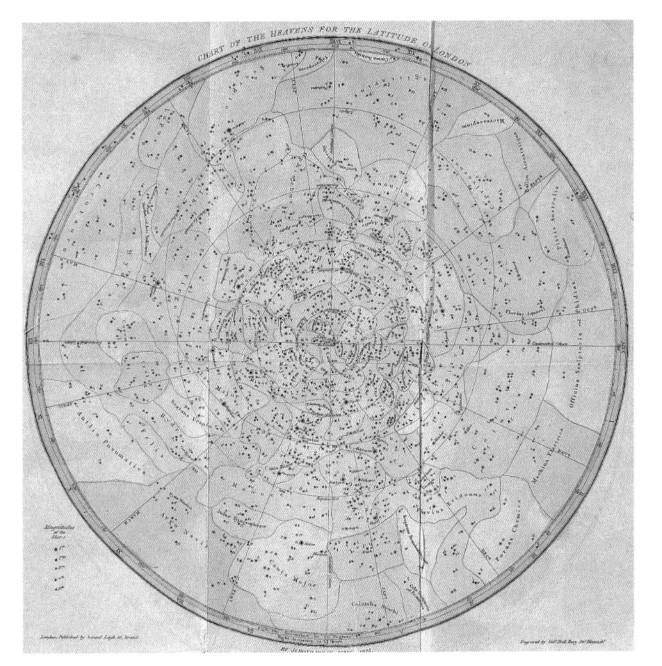

G

Discovering Time

Until the invention of accurate mechanical clocks, astronomical cycles provided the most useful and reliable measures of time. Each day the Sun appears to rise in the east and march across the heavens to set in the west. At noon the Sun is highest in the sky. As it passes overhead toward the west, shadows creep eastward across the ground. Our remotest ancestors surely learned to gauge the time of day by the progress of these shadows.

Before mechanical clocks imposed their regularity upon us, people divided up the day in different ways. Solar time,+ the most important of these, measured the daily motion of the Sun. Popular systems of "seasonal," or unequal, hours divided the day into 12 hours of daylight and 12 of night. This meant that a daylight hour was longer than a nighttime hour in the summer, but shorter in the winter!

Changes of season, plainly linked to heavenly cycles, likewise divide the year. The rising and setting points of the Sun move north for spring and summer, south for autumn and winter (see appendix, p. 144 E). The Sun lags behind the rotating starry heavens so that, as the seasons pass, we see different stars emerging from the eastern horizon+ in the early morning, only to be hidden almost immediately in the glare of the rising Sun (see appendix, p. 144 G). The waxing and waning of the Moon, and its journey through the constellations of the zodiac,+ mark out the months. The stars visible overhead at night also change with the seasons, as a side effect of the Sun's apparent progress through the sky.

All these heavenly cycles have enticed Earth's inhabitants to apply their ingenuity to the measurement of time. Sundials, nocturnals, and astrolabes tell the hour from the position of the Sun or stars. Calendars use the Sun, Moon, or stars to keep track of the date. As people have created instruments to measure time from the heavenly cycles, they have discovered new concepts of time—and new ways to employ them.

Hourly Time

The roots of the word "hour" originally referred to various periods of time, some as long as a season. Ancient Egyptians, likely the first to separate a day into regular intervals, divided daytime and nighttime alike into 12 similar spans, each of which eventually became known as an hour. Beginning about 300 B.C., public sundials displayed this concept of the hour, which slowly spread into common usage throughout the Greek and Roman empires. These hours, sometimes called unequal or seasonal hours⁺ because their duration changes with the seasons, still appeared as curved lines on many Renaissance quadrants and astrolabes. Though equal or uniform hours took root in the 1300s with the invention of public clocks and sand-filled hourglasses, more than five centuries would pass before the hour commonly marked time independently of astronomical or seasonal cycles.

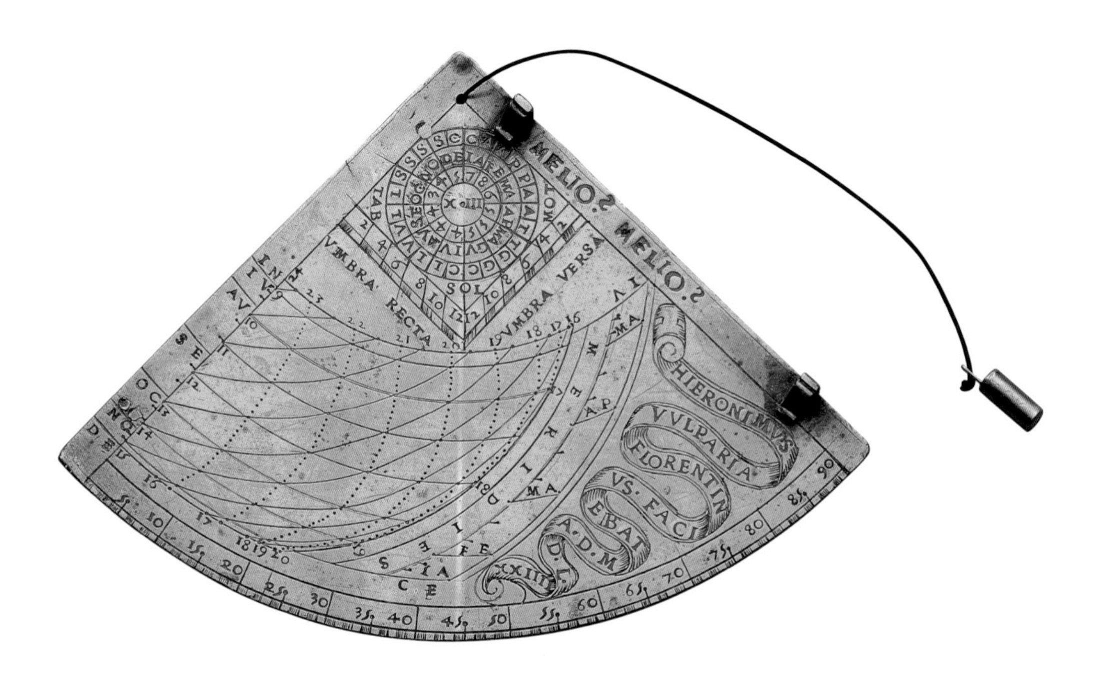

B

D

Finally, read the time by noting the position of the shadow cast on the lines marking the hours, usually labeled with either Roman or Arabic numerals. The circular band on this dial, marked with Arabic numerals, indicates a time just after 12:30 [E].

Some dials also include an adjustment for the date, because the Sun's position in the sky varies greatly over the year and therefore affects the position of the cast shadow. Use of an Augsburg-type dial requires placing the gnomon above the equatorial band during the months between March and September [E], and below it for the rest of the year [F].

E

C

F

Sundials: Unusual Examples

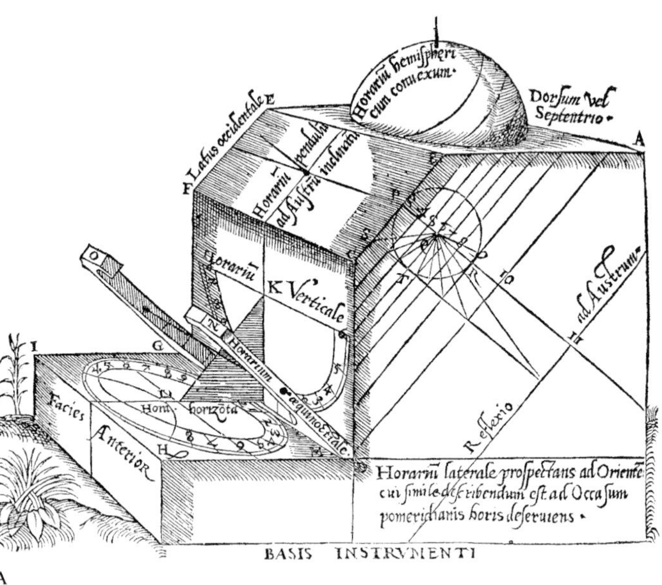

A

B

C

The variety of possible sundial layouts allowed artisans to exercise creativity, ingenuity, and skill. One designer displayed all three traits in a silver polyhedral dial. The product of a plan published by Oronce Fine in 1560 [A], it features an astonishing 29 gnomons[+] for surfaces of different shapes placed at varying angles to the Sun's daily path across the sky [B]!

The ingenious and charming cannon dial proved a popular, audience-pleasing device in the 1800s [C]. When properly oriented, the dial focuses the Sun's rays at noon to ignite a charge placed in a small cannon, thereby providing an audible midday signal.

In devising innovative sundial forms, some creative artisans employed unusual materials requiring careful craftsmanship, yielding dials of great beauty. A fragile glass dial [E] originally served as a stunning window ornament. Beautiful two-panel ivory diptych sundials [D] were luxury items, yet they worked well, providing shadows even in faint light. Larger dials could function as decorative objects, whether made of silver or silvered and gilded brass.

Unusual dials attract our attention with their intriguing features. A pillar dial might include unexpected items, such as a knife and fork [F]. A cruciform dial piques our interest with its shape as well as its contents, the relics of now unknown saints [G].

F

E

D

G

Nocturnals

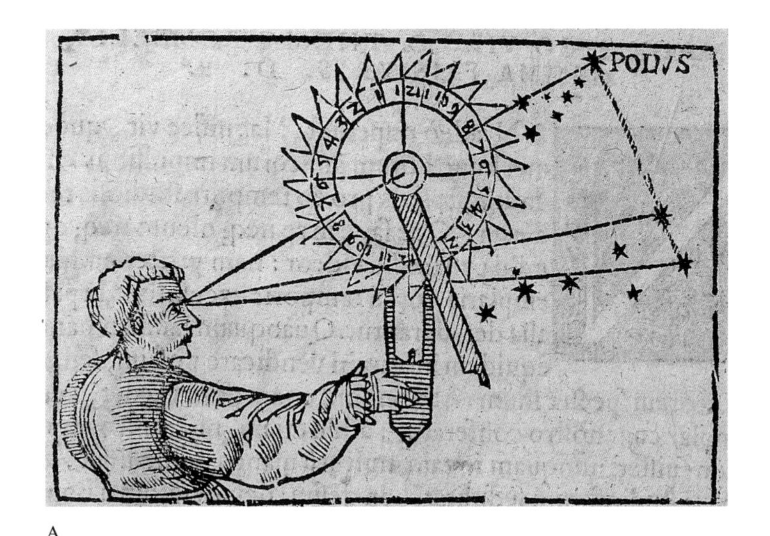

A

B

C

Whereas a sundial traces the daily motion of the Sun, a nocturnal [B] follows the nightly movement of a few select stars. This handy tool uses the sky as the face of a clock, with the Pole Star[+] as its center and another star as its hour hand. Using these coordinates, a nocturnal indicates time to an accuracy of about 15 minutes.

The user holds the nocturnal by a handle attached to the outer of two disks riveted together at their centers [A]. Looking through the central hole at the Pole Star, the user aligns the movable arm of the inner disk with a star in either the Little Dipper or, more usually, the two stars of the Big Dipper that point to the Pole Star. Nocturnals using either star grouping might include such markings as "BOTH BEARS" or "GB" and "LB" for the Great Bear (Ursa Major) and Little Bear (Ursa Minor) constellations [C]. Once aligned, the nocturnal arm indicates the hour of the night.

How does a nocturnal do this? By midnight, each constellation moves to a location in the sky slightly different from its position the previous midnight (see appendix, p. 144 G). The user must therefore adjust the nocturnal to account for this daily change in the celestial clock, a task accomplished by setting the pointer on the nocturnal's inner disk to the appropriate date indicated on the outer disk [D]. The inner disk often includes 24 points, one for each hour, labeled 1 through 12 for morning hours and the same again for the remainder of the day. The outer disk provides information about the

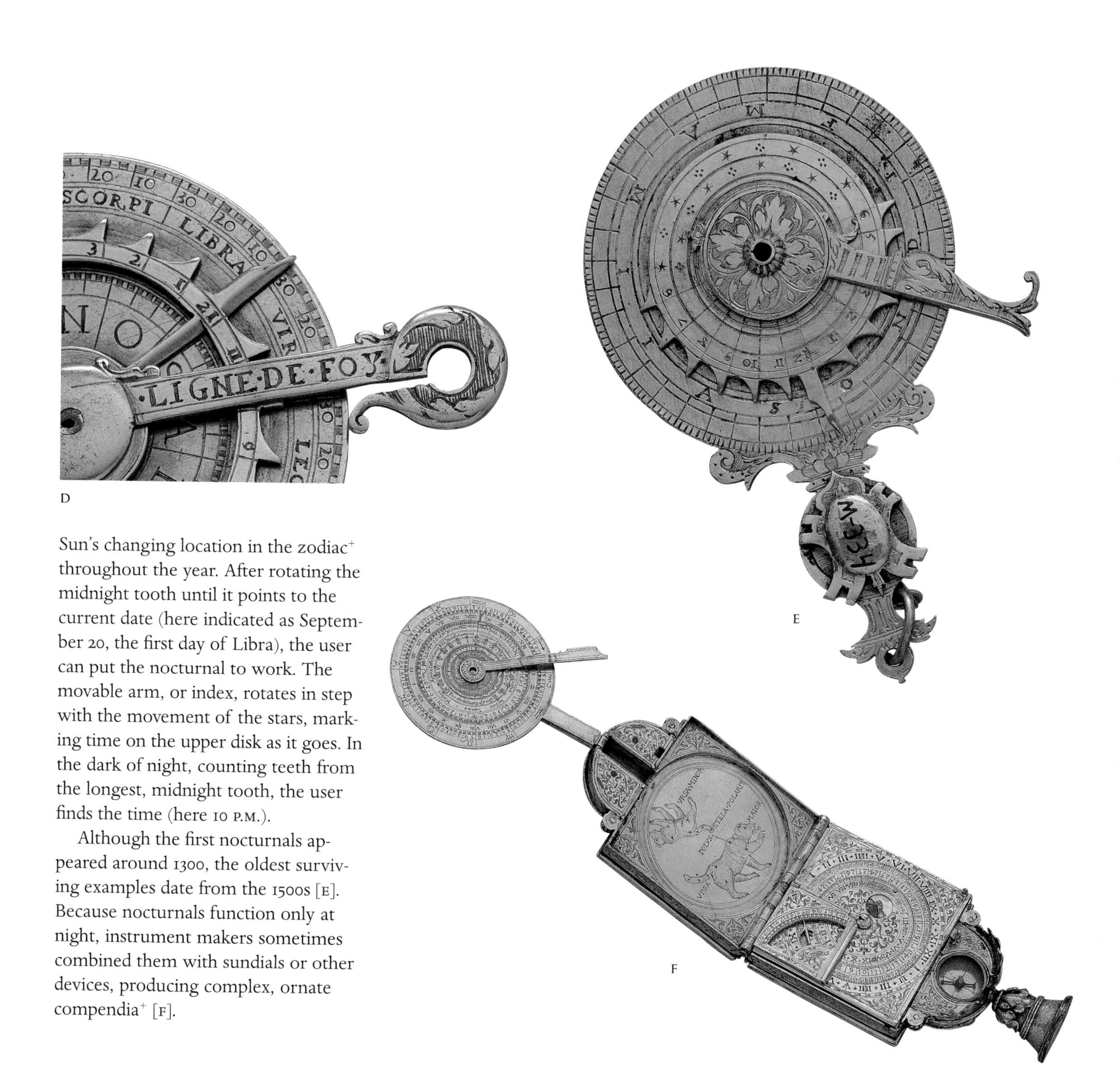

D

E

F

Sun's changing location in the zodiac⁺ throughout the year. After rotating the midnight tooth until it points to the current date (here indicated as September 20, the first day of Libra), the user can put the nocturnal to work. The movable arm, or index, rotates in step with the movement of the stars, marking time on the upper disk as it goes. In the dark of night, counting teeth from the longest, midnight tooth, the user finds the time (here 10 P.M.).

Although the first nocturnals appeared around 1300, the oldest surviving examples date from the 1500s [E]. Because nocturnals function only at night, instrument makers sometimes combined them with sundials or other devices, producing complex, ornate compendia⁺ [F].

Eastern Astrolabes

The astrolabe (see appendix, p. 142 B) is the most sophisticated of the early time-telling instruments. Invented in late antiquity, this all-in-one tool captures the positions of stars and calculates almost anything you want to know about their location in the sky. Knowledge and use of the astrolabe diffused throughout the Islamic world for centuries before reaching Europe about a thousand years ago, entering through what is now Spain.

On the face of an astrolabe [A–C, E], a rotating cutaway disk called the rete+ represents the heavens as they revolve around us. Labeled pointers on the rete indicate bright stars. A prominent ring shows the location of the zodiac,+ the band in the sky where the Sun appears to move in its yearly course. A plate, or tympan,+ inscribed with altitude+ and azimuth+ coordinates (for the particular latitude+ where the astrolabe is being used) lies fixed beneath the rete, visible through its cutaway portions. A curved horizon line on the plate separates the part of the sky visible above the horizon+ from that hidden below the Earth. By turning the precisely crafted rete to the correct position, relative to the plate

C

A

B

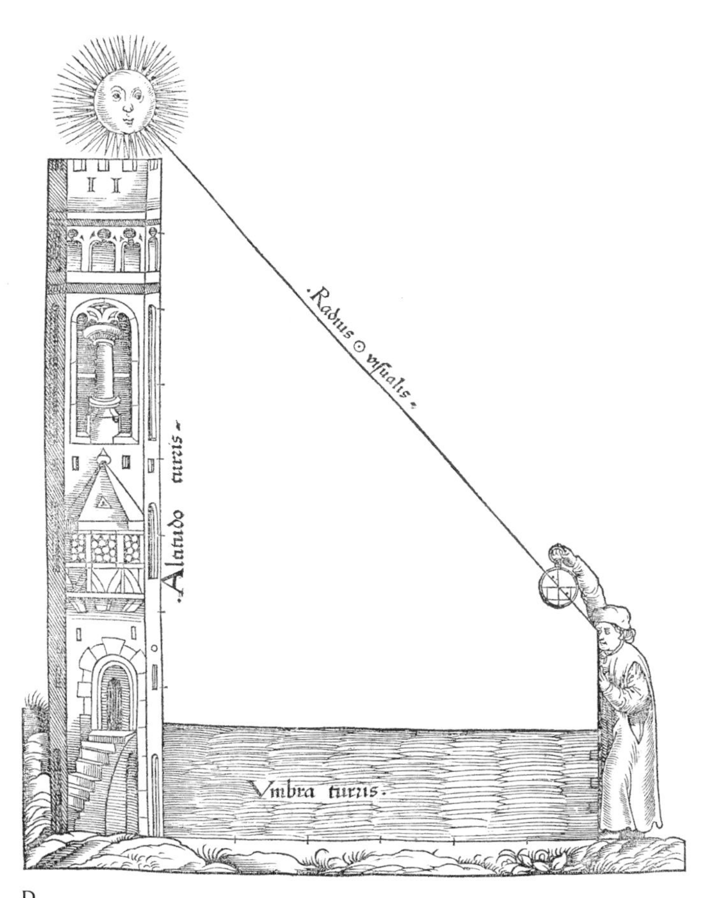

D

E

F

beneath it, you can discern at once the location in the sky of those bright stars to which the rete points. At the same time, you can tell which stars are currently above the horizon and which below.

Because the position of its rete indicates the state of the heavens in their daily rotation, the astrolabe acts as a natural 24-hour clock. At night, a sky watcher suspends it from its ring and observes a bright star with the alidade,† a sighting device usually found on the reverse [D, F]. Looking at the face of the instrument, the user rotates the rete until the pointer for the star rests over the line that marks, on the underlying plate, the altitude observed. The clock is now set, and the Sun's position in the zodiac can be lined up—whether it is above or below the horizon—to tell the time.

For Muslims, an astrolabe served also to indicate the direction of Mecca, the holy city they face when praying. A table gave the azimuth of Mecca from a number of other cities, and a graph on the reverse [F] provided, on any given date, the Sun's altitude when it was at that azimuth. Because of this important cultural function, astrolabes remained in widespread use in the Islamic world well into the 1800s.

Western Astrolabes

A

The Western (European) astrolabe (see appendix, p. 142 B) usually includes two scales wrapped around its back rim, one for the calendar and one for the ecliptic[+] [A, B]. These scales allow easy conversion of the current date into the Sun's location in the ecliptic.

Once the Sun is located in the ecliptic, you can transfer its position to the zodiac ring on the face of the astrolabe [D, F] and find the time, even when the Sun is below the horizon.[+] First set the position of the rete[+] (see pp. 64–65). Then use the rotating rule[+] on the astrolabe's face to line up the Sun's location with the 24-hour time scale around the rim of the face.

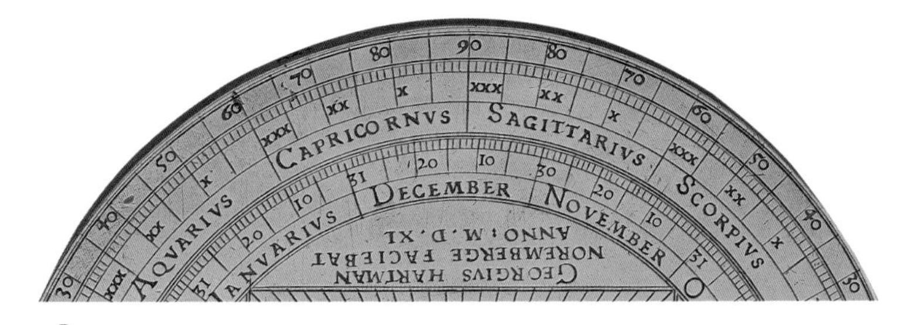

B

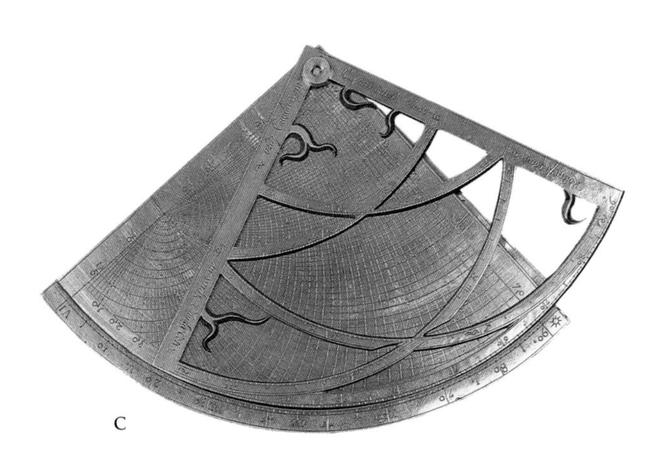

C

D

E

F

G

Western astrolabes are usually inscribed in Latin, the international language of educated people in medieval and Renaissance Europe. Occasionally month names appear on the calendar scale in an everyday language such as French or English [E].

Clever mathematicians and instrument makers squeezed much of an astrolabe's function into smaller, less expensive instruments called astrolabe quadrants [C].

Elegant astrolabes and quadrants were much more than mere timepieces. Indeed they were more than astronomical instruments, for in medieval and Renaissance art an astrolabe signified astronomical (and astrological) wisdom. A fine brass astrolabe was a costly object of great beauty, displaying its owner's learning, wealth, and taste. Less expensive paper versions [G] were printed in books, to be cut out and assembled by scholars. Few of these wood and paper instruments survive today.

Calendrical Time

Slow sky rhythms, the monthly cycle of the Moon and the annual return of the seasons with their characteristic constellations, were even more important to ordinary life than the swiftly passing hours of the day. These slow rhythms brought the tides, the seasons of successful planting and harvest, indeed all the temporal patterns that our ancestors needed to follow. Perpetual calendars once showed the association, almost forgotten today, of the seven days of the week with the seven celestial "wanderers": the two shining luminaries (the Sun and Moon) and the five planets visible to the unaided eye (Mars, Mercury, Jupiter, Venus, Saturn).

DECEMBER.
NOVEMBER.
OCTOBER.
SEPTEMBER.
AUGUSTUS.
JULIUS.
JUNIUS.
MAJUS.
APRIL.
FEBRUARIUS.
JANUARIUS.
WINTER ZONNESTAND.
ZOMER ZONNESTAND.
APOGEUM.
PERIGEUM.
KLIMMENDE-KNOOP
DAALENDE-KNOOP
De Knoopen gaan rond tegen de orde der Zodiaks-tekens in XVIII. Jaaren, 228 dagen, 4 uuren, 52 min. en 52 seconden.
STARKUNDIGE
MAAN-WYZER
EN
ALMANACH.
VERKLARING VAN HET WERKTUIG.
BERICHT VAN HET GEBRUIK.
Nieuwe Maan.
Vierde Octant.
Laatste Quartier.
Derde Octant.
Volle Maan.
Tweede Octant.
Eerste Quartier.
Eerste Octant.
DE MAANDEN DES JAARS.
De ZON in 't Teken van
De lengte der NAGTEN
De OPGANG der ZON te
TYDS-VEREFFENING
JANUARY XXXI.
den 20e
en 8½
15½ en IV
EN 't GETAL der NUMMERDAGEN
deezer Maand: Dan is
de lengte der DAGEN
De ONDERGANG der ZON
VAN X. TOT X. DAGEN.
MIDDELAFSTAND
GROOTSTE NOORDER-BREEDTE
GROOTSTE ZUIDER-BREEDTE
NACHT-EVENING DER LE
HERFST NACHT-EVENING
Noorder declin.
Zuider declin.

Tracking Time with the Sun

B

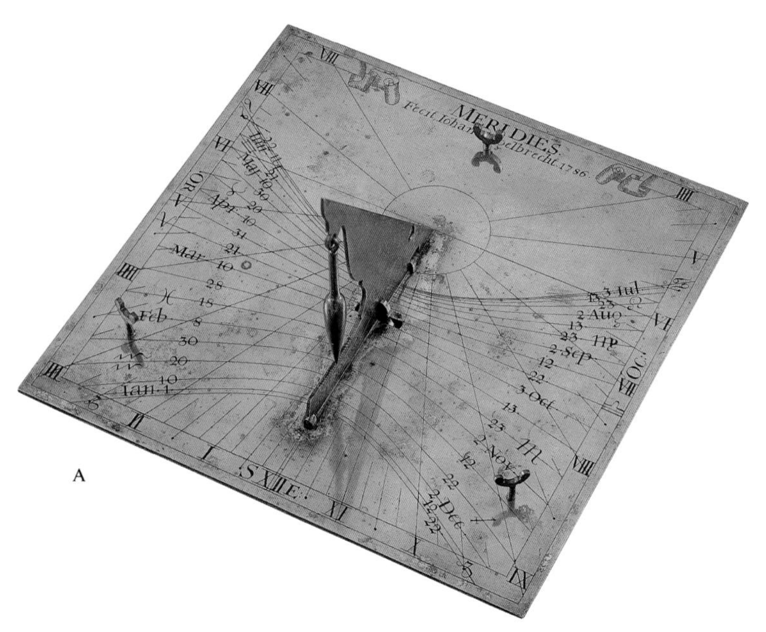

A

The Sun was the first and most prominent timekeeper, showing the time of day by its passage across the sky. Yet it also reveals the time of year. As the seasons pass, the path followed by the Sun each day moves, sliding to the north in spring and summer (for people in the northern hemisphere) and to the south in fall and winter (see appendix, p. 144E). In summer the Sun's path is higher in the sky, so that the Sun beats down more directly and stays above the horizon⁺ longer. In winter the path is lower in the sky, the light more slanted, the days shorter.

Thousands of years ago, people in many civilizations realized that the Sun's daily path slipped north and south in this way, changing the seasons as it did. Important parts of daily life, from praying for a good harvest to simply telling the time of day from shadows, depended on the season of the year [B]. Such activities had to allow for the seasonal changes in the position of the Sun's path.

Sundial layouts ingeniously accounted for seasonal variation. An ordinary sundial from the late 1700s [A], for example,

C

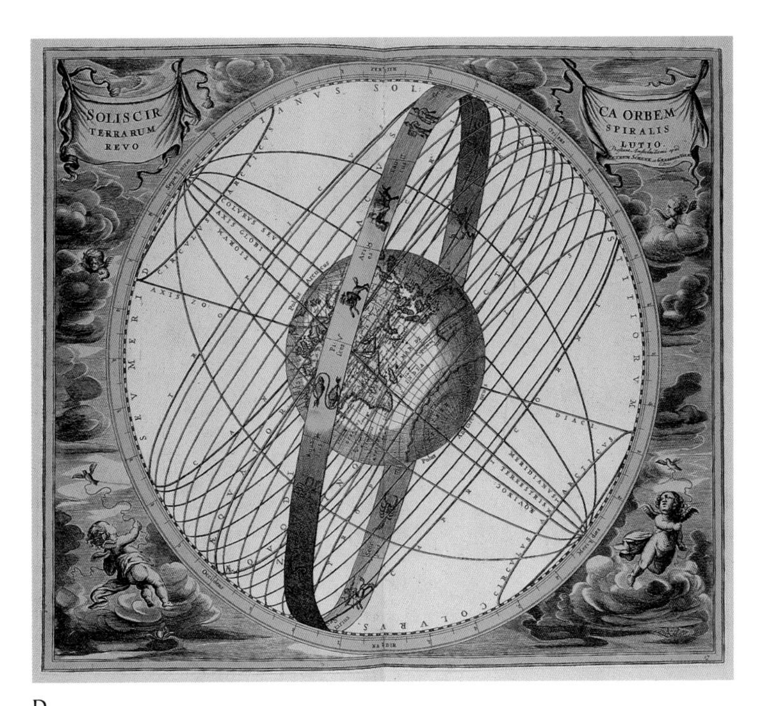

D

indicates the time of day by the angle of the shadow and the time of year by the shadow's length. Sometimes the shadow of a particular point, such as the angle or "elbow" in the gnomon⁺ of an Oughtred-type sundial [C], told simultaneously the time of day and the approximate date.

Changes in the lengths of shadows throughout the year were easy to understand in the Earth-centered cosmos⁺ that medieval Europeans inherited from the Greeks. The Sun moved around the Earth on an inclined circle called the ecliptic,⁺ which carried it north of the equator⁺ in summer, and south in winter. Drawing a picture of this arrangement [D] made it evident that the Sun moved higher in the sky in June than in December. Craftsmen built armillary spheres [E] (see pp. 112–13) to show how the Sun's supposed annual motion around the Earth on the ecliptic shifted its daily path north and south.

E

Tracking Time with the Moon

After the Sun, the Moon serves as our most notable marker of time. The year is structured by its monthly cycle of waxing (growing fatter) and waning (getting thinner) and by its nightly habit of rising and setting about an hour later than the night before. The Moon has served as the basis of calendars for Babylonian, Chinese, Hindu, Jewish, Muslim, and other cultures.

A lunar month captures one cycle of the Moon's phases, beginning when the Moon passes the Sun in the sky, or—since one cannot see the Moon as it passes the Sun—when it has moved far enough beyond the Sun to be visible as a slender crescent above the western horizon,[+] shortly after sunset. For a couple of weeks, the shape of the Moon grows fatter, and it stays above the horizon longer and longer each night. Halfway through the month, the Moon is full, a complete globe that rises at sunset and sets at sunrise. For the last half of the month, the Moon rises after dark and is still above the horizon at sunrise, meanwhile growing thinner each day. At the end of the month, the Moon rises shortly before the Sun, appearing as a thin crescent that is soon lost in the glare of the rising Sun.

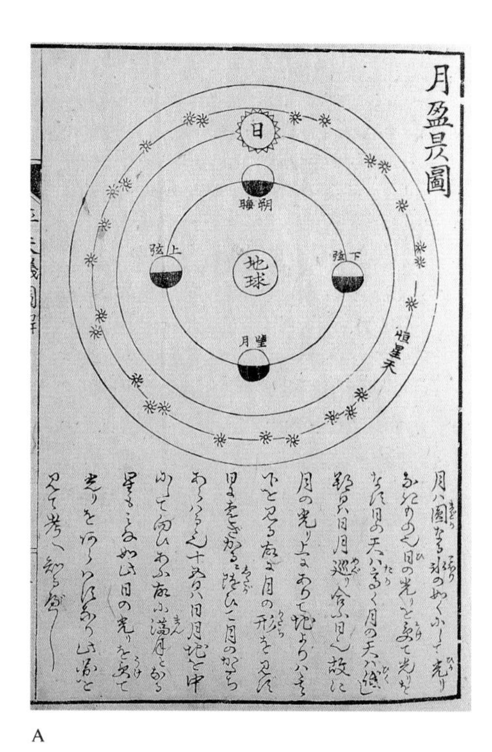

A

B

C

D

Long ago, sky watchers correctly interpreted the cycle of the Moon's waxing, as it moves away from the Sun, and waning, as it approaches the Sun. The Moon shines with reflected sunlight, and we see only the part of it that is both illuminated (facing the Sun) and visible (facing the Earth) [A, B]. Lunar eclipses confirm this explanation, because they do not happen unless the Earth is directly between the Sun and the Moon (see p. 33 E).

The lunar cycle provided a popular subject for volvelles:+ movable constructions, usually of brass [D, E] or paper [C], that showed the changing phases. These devices, the interactive projects of their time, allowed people to physically manipulate a simple model and re-create what they saw in the sky.

E

Tracking Time with the Stars

B

A

Every season has its constellations. Until electric lighting gathered us indoors and dulled our night vision, people could always glance at the stars and tell the season of the year. How was this possible? After all, the stars do not move much in the course of a year. For most of history, indeed, they were called fixed stars, because their positions relative to one another did not visibly change.

The stars seem to move slowly westward from night to night, however, because of the apparent motion of the Sun (see appendix, p. 144 G). The Sun moves through the constellations of the zodiac^{+} [B], traveling from west to east, by about twice its own width each day. You cannot see the stars at all when the Sun is nearby, so this motion is noticeable only indirectly, at sunrise and sunset. The constellations^{+} visible at sunset or sunrise change with the seasons [A].

The Hunter and the Bull (Orion and Taurus) are winter constellations, and Sirius a winter star, because in winter the Earth has moved between them and the Sun. As a result, they are all visible above the horizon^{+} at night, when the Earth has turned us away from the Sun. The Lyre (Lyra) is a summer constellation, and the bright stars Vega, Altair, and Deneb are known as the "summer triangle," [C] because the Sun has moved far from them in summer and they are high in the sky after nightfall. Because the positions of the stars

C

depend on the time of year as well as the time of night, you must set the date on a nocturnal [D] before you can use the stars to tell time (see pp. 62–63).

On the plate of an astrolabe [E], the curving horizon line divides heavenly bodies above the horizon from those hidden below it. As the Sun moves around the ecliptic⁺ ring in the course of the year, it is the stars whose pointers are located across that ring from the Sun's position that will be above the horizon, and visible, after the Sun has set.

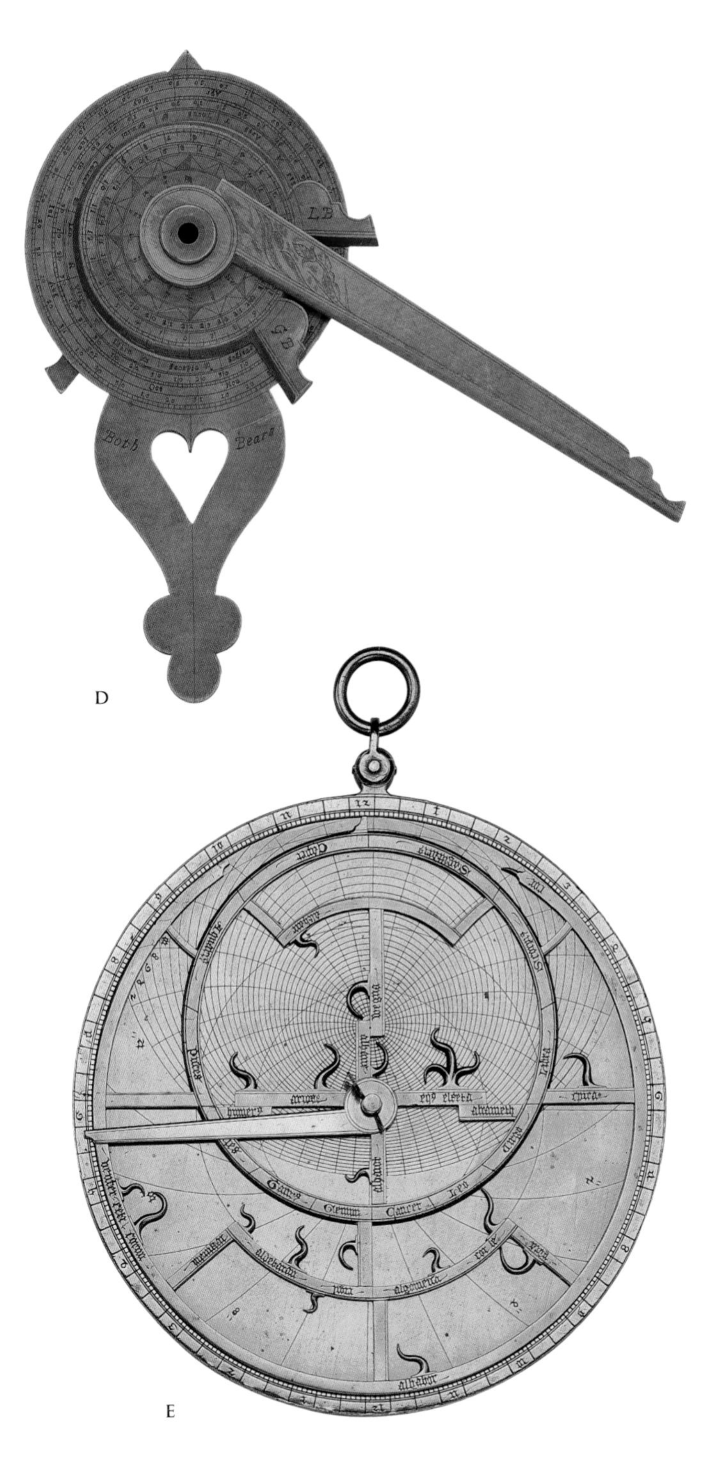

D

E

Draco
Bootes
Corona Sep
Major
Libra
Scorpio
Pavo
AME
NOR
THE
SOUTH
SEA
OCE
Nova Francia
Newfound Land
New England
Mary land
Virginia
Florida
Hispania Nova
Mexico
Sinus Mexico
Bermuda
Cuba
Jamaica
Barbadas
Nova Granada
Guiana
Amazons
Peru
Brasi lia
Circle
Capricorn

Understanding the Earth

As people discovered the universe around them, they sought to understand not only the heavens but the Earth as well. Often they imagined a relationship between the two. Astrology explicitly linked the heavens to Earth by attempting to predict and explain terrestrial events based on the positions of celestial bodies at certain times.

Since ancient Greek times, geographers have known that the Earth is round, though they envisioned it nested in the center of a sphere of planets and stars. As curious individuals began to explore the globe, often guided by the stars, they desired to depict the relationship between land and ocean, measure the distances between cities, and locate geographical features. Cartographers recorded such information on globes and maps so that others could make use of it.

But measuring and mapping did not happen only on a grand scale. Surveyors+ developed techniques to measure (and subsequently to map) the length of roads, the size of plots of land, and the height of mountains. To speed the process of creating maps and solving mathematical problems, instrument makers produced drawing and calculating tools. Uniform standards of measurement were introduced, though at first they varied from city to city and from country to country.

The development of accurate, mechanical methods of determining time ended our reliance on knowing the hours of daylight and darkness in particular seasons. Now able to measure time uniformly and precisely, we finally could determine longitude+ consistently and correctly.

The Earth Mapped

To record their surroundings, many humans have mapped the world around them. Globes represent the entire planet, while flat maps depict anything from relatively tiny areas, such as a town, to the entire Earth. Many people could sketch a rough outline map of a small area, but mapping broad ocean expanses and interior geographical features requires considerable skill. After instrument makers developed and refined devices to measure distance, direction, and time, geographers could create maps and globes that more accurately represented the Earth's features.

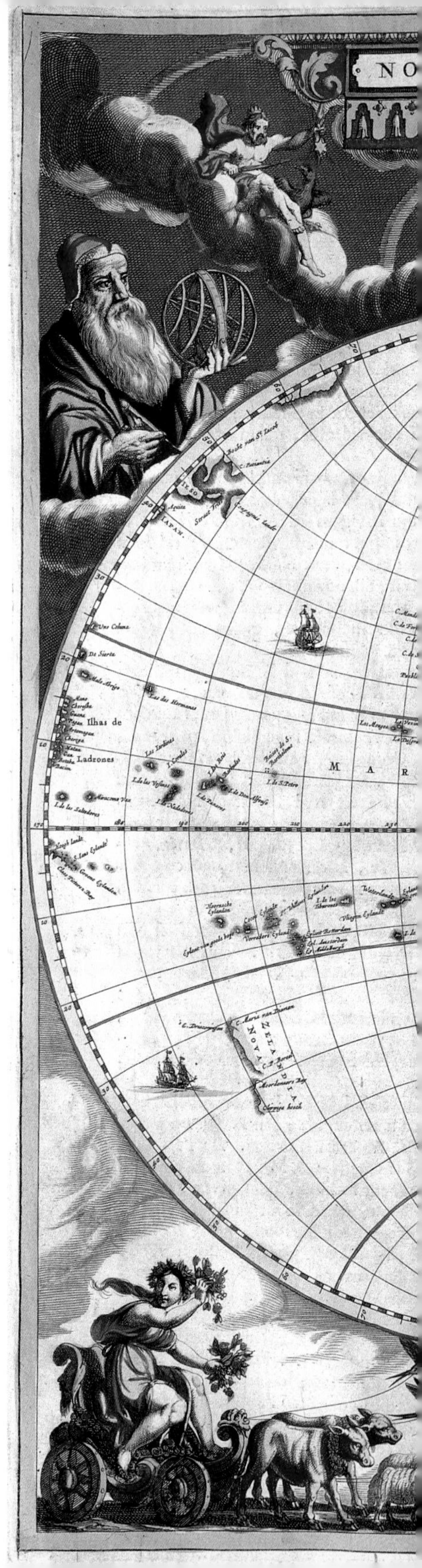

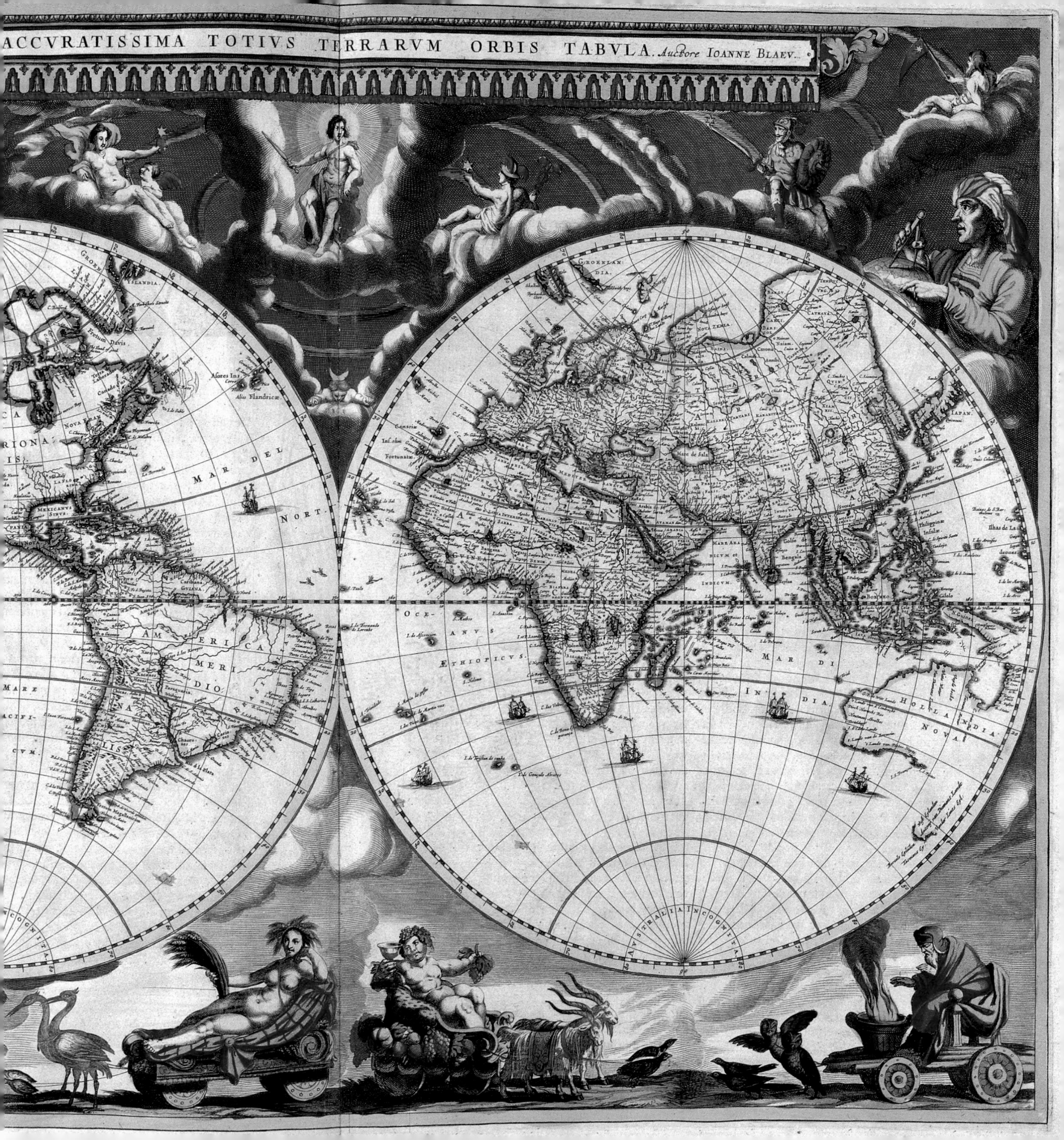

ACCVRATISSIMA TOTIVS TERRARVM ORBIS TABVLA. Auctore IOANNE BLAEV.
MAR DEL NORT.
AMERICA MERIDIONALIS
MARE PACIFICVM
OCEANVS ÆTHIOPICVS
MAR DI INDIA
HOLLANDIA NOVA
AFRICA
ASIA
AVSTRALIA INCOGNITA

Terrestrial Mapping

Until the late 1400s, a European scholar's concept of the Earth greatly resembled that of the ancient Greeks as described in Ptolemy's *Geographia,* written in the second century A.D. Their known world consisted only of Europe and parts of Asia and Africa [C]. Indeed, most people had little knowledge of places outside their immediate locale until printed maps began to appear in Europe in the 1470s.

During the age of discovery, explorers such as Christopher Columbus, Bartholomeu Díaz, and Ferdinand Magellan radically changed our knowledge of the world. In the early 1500s, cartographers began to produce maps that indicated the new lands discovered by these explorers. By the mid-1500s, popular cosmography+ texts invariably included a world map (see p. 39 D).

Some maps are little more than sketches outlining the general shapes of the continents [B]. Others are more detailed, complete with place names, rivers, and mountains (see p. 79). Cartographers enlivened their maps with decorative additions such as ships and sea monsters in the oceans and naturalistically drawn animals and native peoples on land.

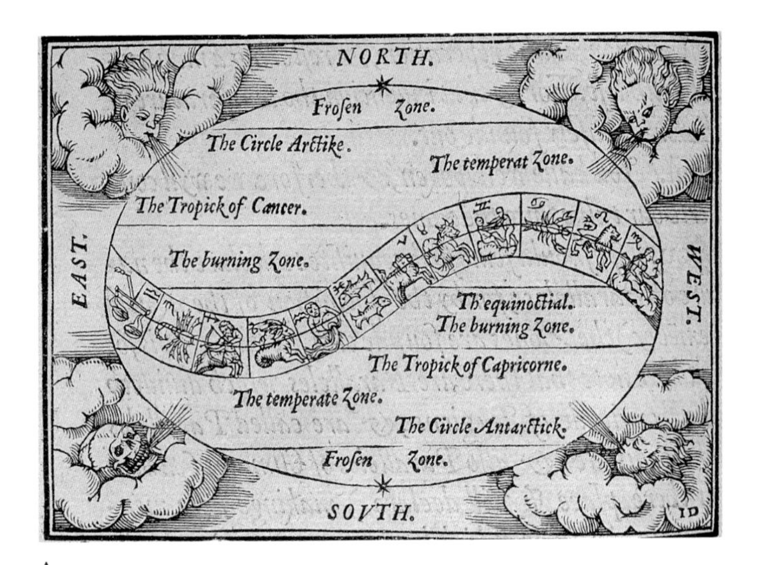

A

B

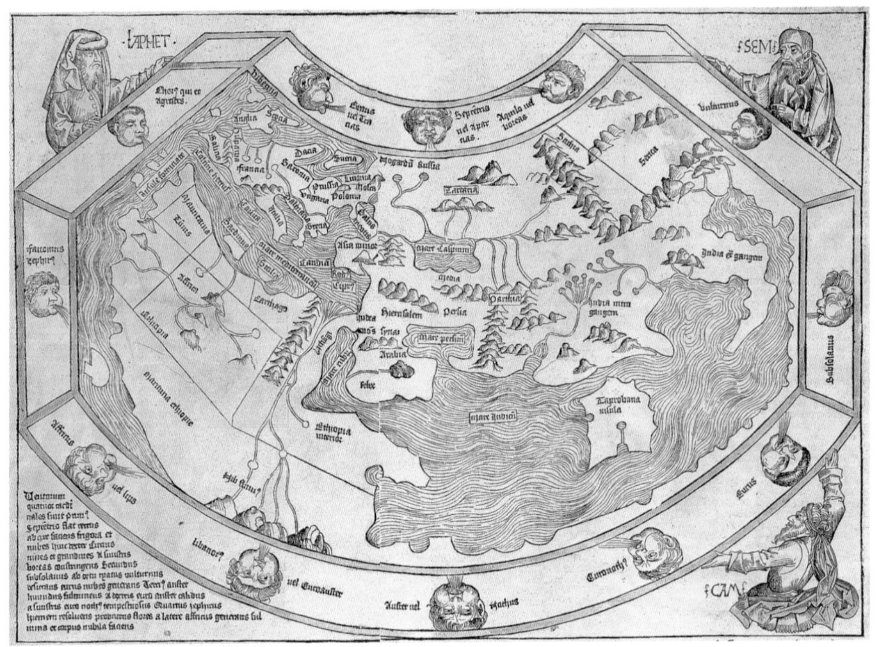

C

D

E

Still other maps depict not land masses but the climatic zones delineated by the polar circles,+ tropics,+ and equator+ [A].

Certain sundials and compendia+ sometimes include regional maps, ostensibly to aid the traveler who might have carried such an instrument [D]. Armchair travelers took great delight in maps, gazing upon all the varied places in the world and imagining what they might be like.

Globes constitute another form of terrestrial map. In addition to representing the spherical nature of our planet, they depict its surface features [F]. But some globes are not just maps of the Earth [E]. With the addition of a horizon+ ring, a movable meridian+ ring, and an hour circle, these instruments demonstrate the rising and setting of the Sun. Though instrument makers constructed early globes so that the Earth rotated about its axis, people still believed that the Sun revolved around the stationary Earth.

F

Eclipse Maps

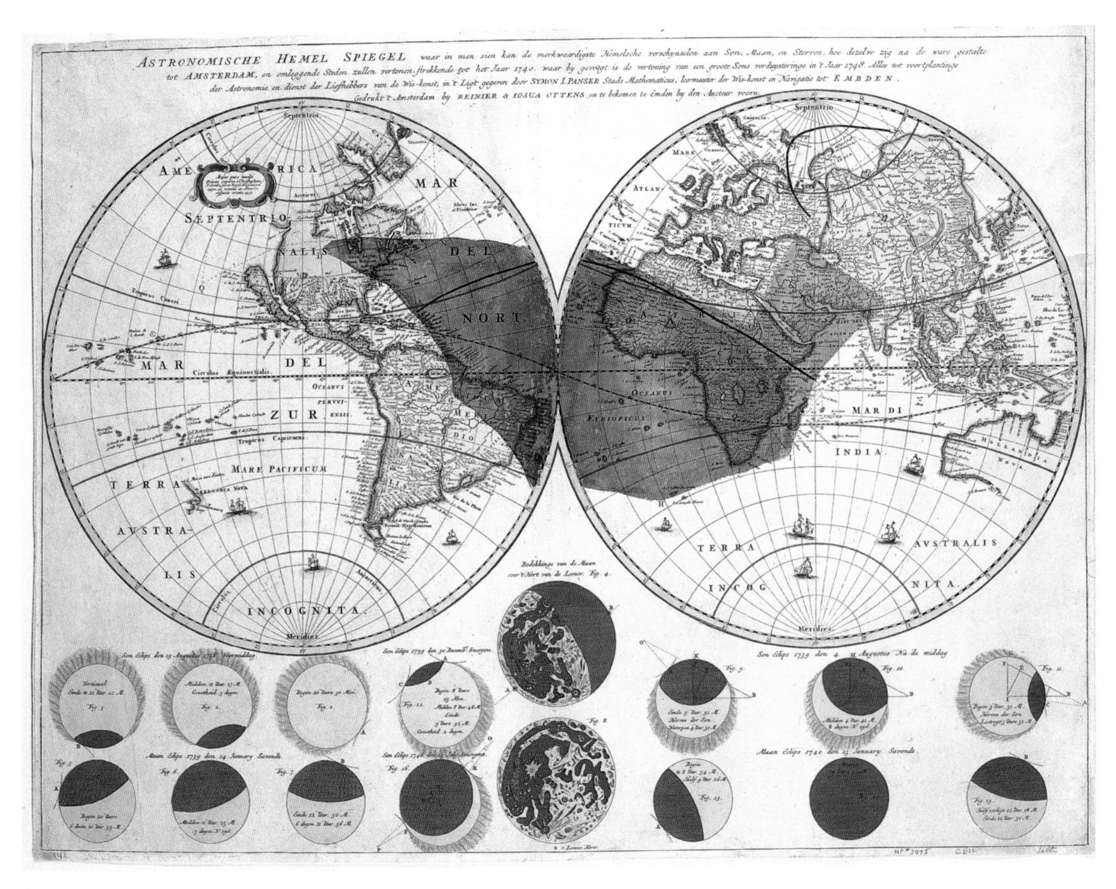

A

Blending terrestrial cartography and astronomy, eclipse maps show the extent of the shadow cast onto the Earth during a solar eclipse. A shaded area superimposed onto a geographical map mimics the darkening effect of an eclipse.

Such maps came into being during the 1700s as astronomers learned to predict the area that would be affected by a solar eclipse. The famed astronomer Edmond Halley created the first eclipse map of this sort, almost perfectly predicting the path of the eclipse of 1715 [B]. Cartographers drafted maps before an eclipse occurred, to predict the path, or afterwards, to document its actual path.

Some maps showed the paths of several eclipses that took place on different dates [C]. An overview of various eclipses allowed the viewer to compare the locations of the shadows.

Of use not only to astronomers, these maps could be easily read by lay people seeking details about a particular eclipse. Astronomers often added diagrams of how eclipses happen or circular charts showing the different levels of totality around the world [A]. Separate broadsides⁺ also contained this information [D]. Such additions educated the general public about the astronomy behind eclipses.

C

Surveyors and cartographers considered their drawing and calculating tools indispensable to their crafts. With these devices, they transformed their measurements and observations into straight lines of particular lengths, either parallel or at specified angles to each other. The arms of a divider, sometimes called a compass [E], can be set to a desired opening, determined either by direct measurement appropriately scaled down or by a scale marked on a sector. Like slide rules, sectors include scales of various mathematical functions; they also provide additional information, usually relating to gunnery or other military needs. Some even list specifications for building fortifications [F].

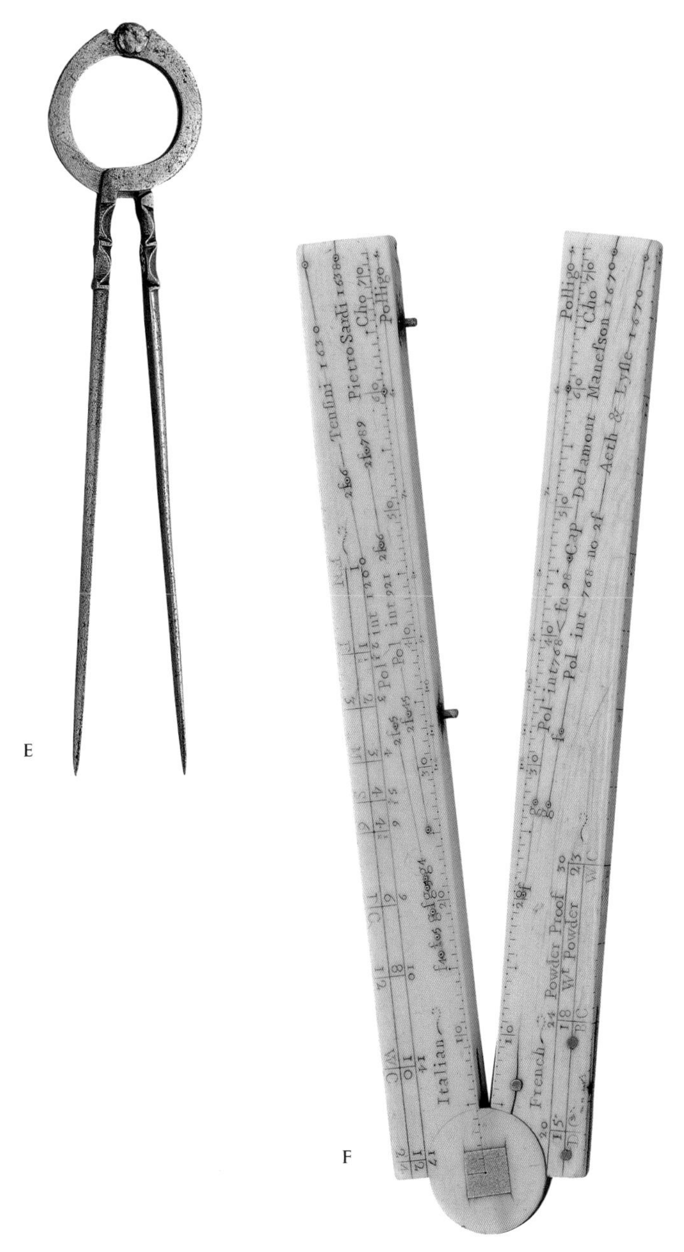

E

F

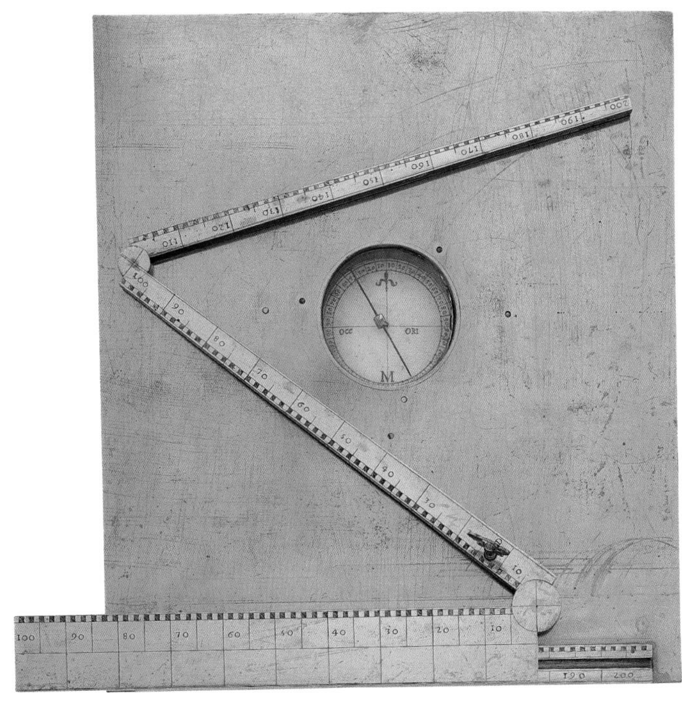

D

Mechanical Time

The unequal hours⁺ of solar time⁺ ebb and flow with the seasons, with longer daylight hours in the summer and shorter daylight hours in the winter (see pp. 53–54). But the march of mechanical time ignores such natural cycles, keeping in step only with the swinging of a pendulum and the turning of gears. Clock makers built public clocks throughout Europe in the mid-1300s. Two hundred years later, they produced monumental public clocks, whose large and numerous faces symbolized universal stability and order, their operation a reminder of the inexorable passage of time.

From Solar to Mechanical Time

Until the 1800s, mechanical clocks remained quite unreliable, requiring regular adjustment against a sundial. A sundial indicates local solar time,[+] based on the current position of the Sun in the sky (see pp. 58–59). Because the Sun's movement across the sky changes speed slightly over the course of the year, the time indicated by a sundial varies from that indicated by a reliable mechanical clock, which runs at a constant rate. The difference between solar time and mechanical clock time,[+] known as the equation of time (see appendix, p. 144 F), shows that a sundial runs "behind" a clock by nearly 15 minutes in February, and "ahead" by more than that in October.

To help its owner keep time more accurately or to reset a clock or watch, some dials include the equation of time, whether printed [A] or incorporated into the dial itself. A few dials, such as some equatorial dials [B] and heliochronometers [E], feature a mechanism that translates the Sun's movement onto a clock face.

In the early 1800s, people still considered time in terms of the Sun's position or other natural cycles, as evidenced by the scenes found on some common clocks of the period [C]. Then, as now, in towns separated by a dozen miles east to west, the moment at which the Sun reaches its highest point at noon differs by about a minute. Towns more distant show

A

B

C

D

a greater discrepancy on their sundials, but with no way to communicate quickly over large distances, people were little troubled by the variance.

This changed in the 1830s with the development of the railway, the foremost cause of the sundial's demise. Passengers would have found it too cumbersome and complicated to consult a railway timetable listing local solar times. Instead, timetables used a common or standard time, which in Britain was based on the local time at the Royal Greenwich Observatory. By the 1850s, the telegraph synchronized British railway clocks to Greenwich time and enabled rapid communication over the vast distances of the United States. By 1900, North America, much of Europe, and various regions around the world had made the transition from local solar time to standard time zones. Increasingly, an organized network of mechanical clocks, not sunlight or other natural cycles, regulated the events of the day [D].

E

Clocks and Watches

To mark the passage of time, people in the 1300s began to use the sand-filled hourglass (see pp. 54–55) and the mechanical clock, fitted with an escapement⁺ to regulate its movements and bells to announce the hours. The monumental public clock, with its display of astronomical information, became part of a city's identity and even its symbol in art. Early private clocks usually featured an hour hand but no minute hand, a feature [B] rare even after the improvement of the pendulum clock by Christiaan Huygens in the 1650s [A].

Because early clocks needed a lofty perch for their falling weights, only the residences of royalty and the wealthy could house them. Clocks became available to a larger market in the mid-1400s, as clock makers found ways to make them smaller and more portable. When continued miniaturization and portability produced the pocket watch, by 1500, time entered the private arena, encouraging individualism and personal productivity. Many early watches indicated only the hour [C], with later ones adding minutes as well as astronomical information such as the lunar phase [D].

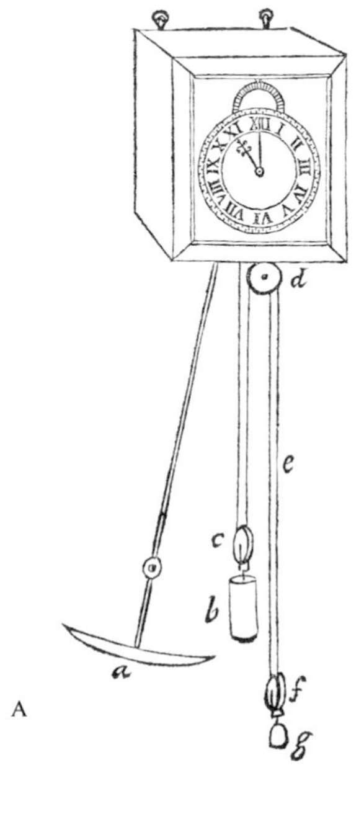

A

B

C

D

The Longitude Problem

Anyone can find local latitude⁺ simply by measuring the altitude⁺ of the Pole Star⁺ (see p. 39). But finding one's longitude—degrees east or west from a reference point, say that of the Royal Greenwich Observatory—proves considerably more difficult. Every 24 hours, the Earth moves 360 degrees in longitude; every hour corresponds to 15 degrees, every 4 minutes to 1 degree. Near the equator, where each degree measures more than 60 miles, a timepiece accurate to only 4 minutes per day could produce a longitude error of more than 60 miles, each day! Because shipboard clocks in the 1600s could not even approach that level of timekeeping, navigators could not accurately

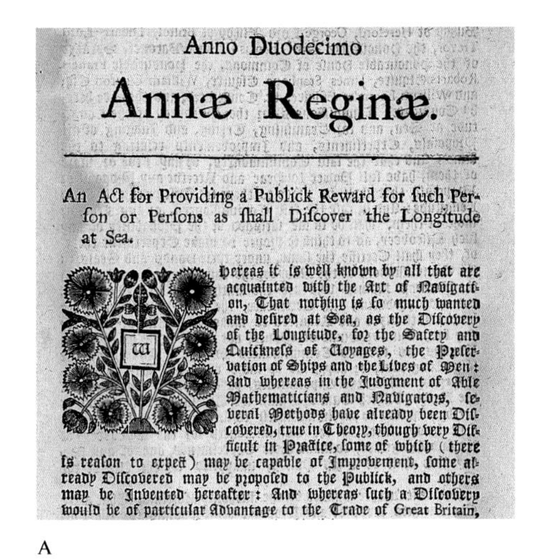

Anno Duodecimo

Annæ Reginæ.

An Act for Providing a Publick Reward for ſuch Perſon or Perſons as ſhall Diſcover the Longitude at Sea.

Whereas it is well known by all that are acquainted with the Art of Navigation, That nothing is so much wanted and desired at Sea, as the Discovery of the Longitude, for the Safety and Quickness of Voyages, the Preservation of Ships and the Lives of Men: And whereas in the Judgment of Able Mathematicians and Navigators, several Methods have already been Discovered, true in Theory, though very Difficult in Practice, some of which (there is reason to expect) may be capable of Improvement, some already Discovered may be proposed to the Publick, and others may be Invented hereafter: And whereas such a Discovery would be of particular Advantage to the Trade of Great Britain,

A

determine the difference between local and Greenwich time and, therefore, between local and Greenwich longitude.

Early in the 1700s, the problem of finding the correct longitude of a ship at sea became so pressing that the British Parliament offered a fortune to anyone who could discover a reliable method [A]. Edmond Halley's strategy of mapping global variations between magnetic and geographic north failed [B]. Charting the Moon's position against the starry background, the so-called lunar distance method, required only a simple conversion to yield a longitude difference, and it proved fairly reliable for determining the time difference between local noon and Greenwich noon. In the end, though, the chronometer [C], which keeps time to within parts of a second per day, provided the most accessible and accurate solution. It eventually won its inventor, John Harrison, the prize money and brought prestige to those who made or owned such an instrument.

B

C

Portents and Predictions

A blend of astronomy and mythology, astrology once played a prominent role in daily life. Though it is now usually dismissed as frivolous, people once considered it a science equal to astronomy. In fact, the terms "astrology" and "astronomy" were almost interchangeable prior to the 1600s. The diverse cultures of Europe, Arabia, China, India, and Egypt all practiced different types of astrological divination. The astrology we know today started in ancient Mesopotamia. Astronomers in the ancient Greek, Roman, and Islamic worlds refined the discipline. People in medieval and Renaissance Europe employed astrology often to explain and predict various phenomena. By the 1700s, however, interest in astrology had waned in scientific circles.

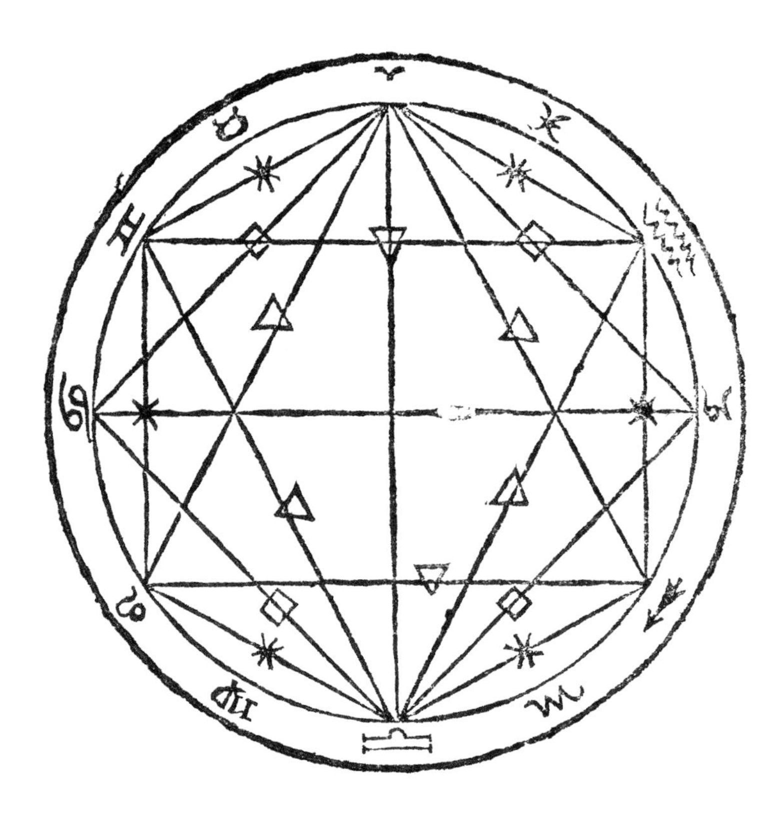

IDEA DELL'
VNIUERSO
CANCER GEMINI
BOREAS AQVILO
CAPRICORNVS
SAGITTARIVS
AFRICVS
All'Ill.mo Signore
Il Sig. Abbate Sebastiano Venier
Patritio Veneto

Astrology and Daily Life

For thousands of years, people looked to the sky to help explain happenings on Earth and to reveal details about their lives [A]. In consultation with astrologers, they found solace or meaning in the answers and predictions gleaned from heavenly divination. Although without scientific grounds, astrology derives from astronomical calculations—the positions of the planets, Sun, and Moon with respect to one another and the vernal equinox—or the interpretation of astronomical events.

Astrologers cast horoscopes⁺ to reveal such information as future manifestations of a newborn's personality [B] or the best day to celebrate a wedding. Farmers consulted almanacs⁺ to determine the best day and time to plant different crops. Astrological signs corresponded to different parts of the body [D], and doctors used astrology to determine the nature of a person's illness and how to treat it.

A

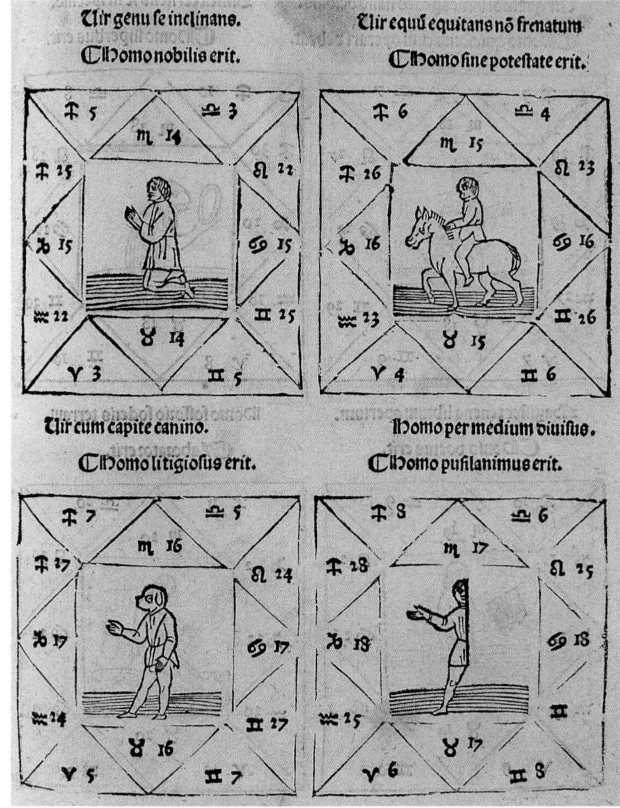

B

Ihr Betrachter und Beobachter der täglich-neuen Welt-Begebenheiten/
Stehet/ Schauet/ und Erstaunet/
Uber einer abermahl in der Lufft aufgesteckten Göttlichen Zorn-Fackel/
Welche Anlaß gibet zu Erneuerung und Verdoppelung deß letztlich entworffenen
Cometen-Memorials und Zeigers an der Göttlichen Zorn-Uhr/
Da über den vor dritthalb Jahren im Christmonat deß 1680sten und im Jenner deß 1681. Jahrs erschienenen erschröcklichen Cometen A. der Statt Augspurg eine stetswehrende Ab- und Vorbildung ihres Forcht-erweckenden Anblicks ist gezeiget; und Anno 1682. im Augst- und Herbst-Monat ein neuer entsetzlicher Straff-Prediger B. nach Gottes Willen beygefüget worden. Nun aber der dritte C. im Julio und Augusto dises 1683. Jahrs/ mitten unter denen grausamen Kriegesflamen/ uns hochnachdencklich von dem erzörneten GOtt für Augen gestellet wird.
Lachet ihr nun hierüber ihr sichere Welt-Kinder?
Verachtet ihr es/ ihr ruchlose Epikurer?
Und verharret ihr in eurer Boßheit/ ihr verstockte Sünder?
So überleget und erweget doch Ihr/ ihr wohl-geartete und rechtschaffene Christen-Gemüther/
Was sint der Zeit hero da und dort/ hin und wieder/ für Jamer und Noth/ Seuche und Tod/ Uberschwemung und Fluth/ Verderben und Gluth/ schädliche Kriege und Zerstörungen/ hinterlistige Anschläge und Zerrüttungen/ ja unzehliche böse Stücke/ arge Tücke/ und grosse Unglücke/ auf jenen Cometen (nicht eben als einen Wircker und Ursacher/ sondern vilmehr als einen Vorbotten und Warner) erfolget/ entstanden und ergangen;
Seit dahero geflissen euer seits so vil möglich der Gerechtigkeit Gottes in die Ruthen zu fallen/ damit seine Barmhertzigkeit das noch hinterstellige/ weit hinaus böß-aussehende/ androhende Ubel/ abwenden/ und alle dergleichen Unglücks-Posten ferner von uns gehen lassen möge.
Verhindert die scharffe Würckungen diser Zorn-Zeichen durch Busse und Besserung;
Seit getrost im Creutz und Leiden/ beständig in Gedult und Hoffnung/
Gehet/ Wachet/ und Bettet.
Gedruckt zu Augspurg/ bey Johann Jacob Schönigk.

C

Astrologers also interpreted celestial happenings such as eclipses, comets, novae,+ and planetary conjunctions+ to predict or explain events ranging from the outcome of wars to the deaths of rulers. They often published accounts of these events on broadsides+ [C] and in small pamphlets [E].

Popularly thought to be portents of death, war, plague, pestilence, storms, and political upheaval, comets played a significant role in astrology and in other forms of prognostication. The constellations a comet crossed as it traveled through the sky, its magnitude,+ the length and direction of its tail, and its color contributed to various astrological interpretations.

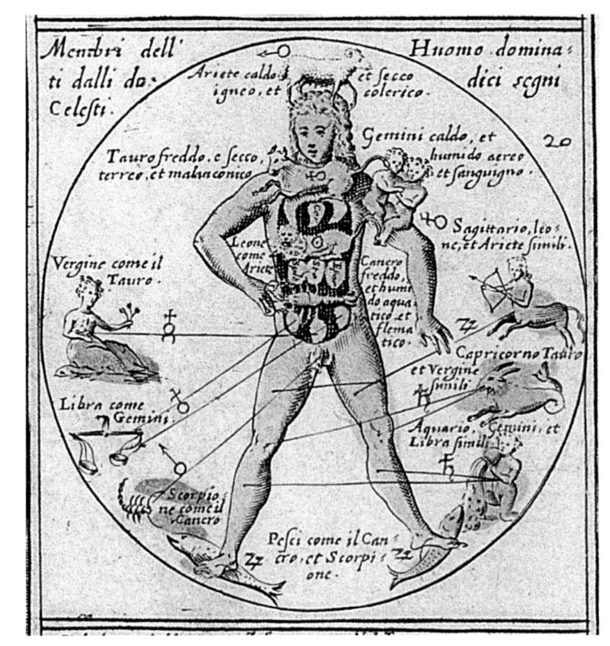

D

Was ein Comet sey: wo her er kome/ vñ seinen vrsprung habe/ von vnderscheidung/ vñ in was form vñ gestalt sye erscheynen. Auch von irer bedeütung/ mit anzeygung etlicher historien vnd geschichten/ so denen Cometen nach geuolgt/ vnd sonderlich von dem Cometen erschinen im Weynmonat des xxxij. jars. Durch Nicolaum Prucknerum beschriben.

E

Casting Horoscopes

Whenever people believe that astrologers can tell them what to expect from life, the business of casting horoscopes⁺ flourishes. The drawing up of an astrological chart could be a work of careful analysis and tedious calculation (particularly if the subject was a prince!), but an astrologer needed quicker methods when less was at stake.

An astrolabe, although a serious astronomer's tool, is almost ideally suited for the rough-and-ready calculation of astrological charts. It does not help in finding the positions of planets in the zodiac⁺—for that you need a set of tables or a calculating device called an equatorium. The astrolabe is, however, an ideal tool for locating those planetary positions among the "mundane houses," the 12 all-important divisions of the zodiac that originate, each moment, from the ascendant,⁺ where the ecliptic⁺ rises from the eastern horizon.⁺ Most European astrolabes bear lines specially engraved to show the boundaries (called cusps) of each mundane house [A]. Celestial objects that happened to be in each of these 12

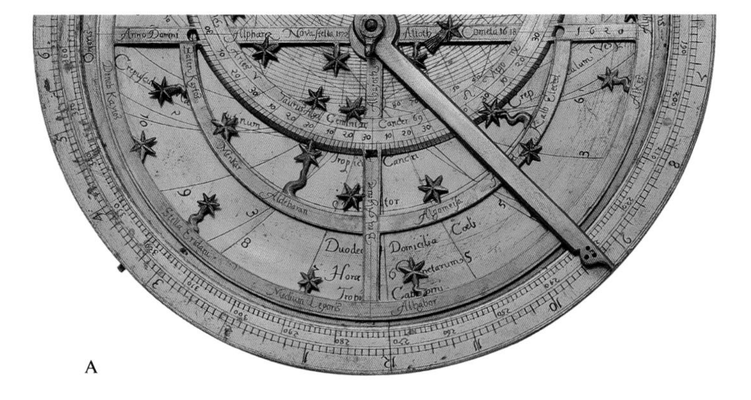

A

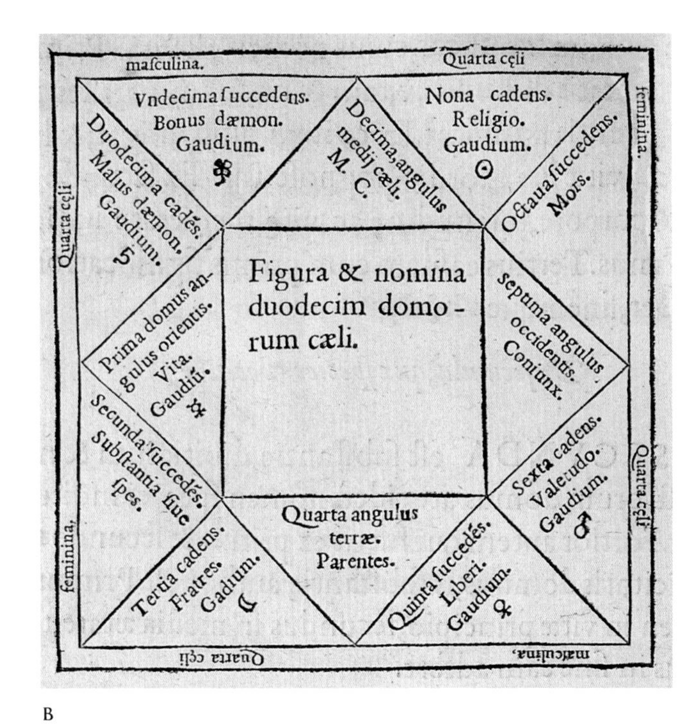

B

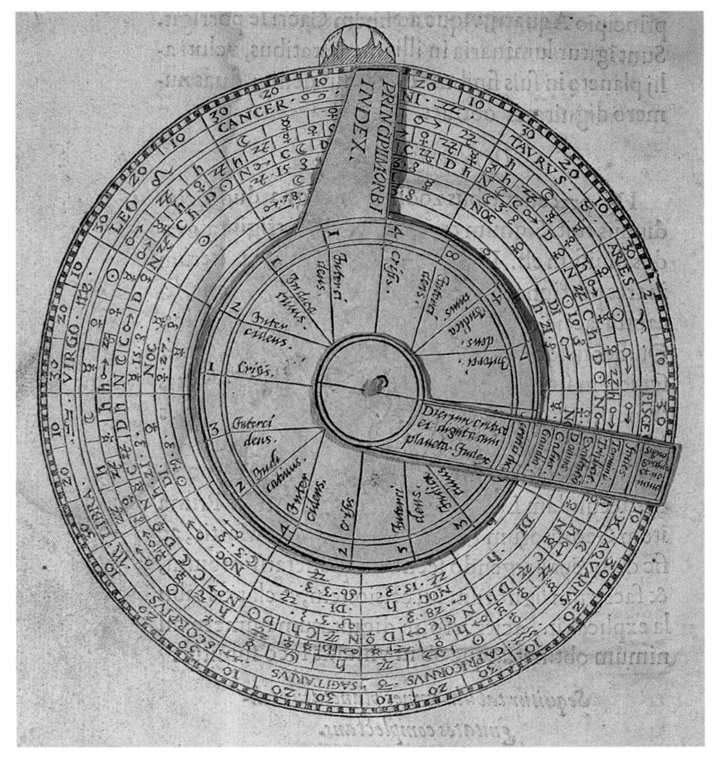

C

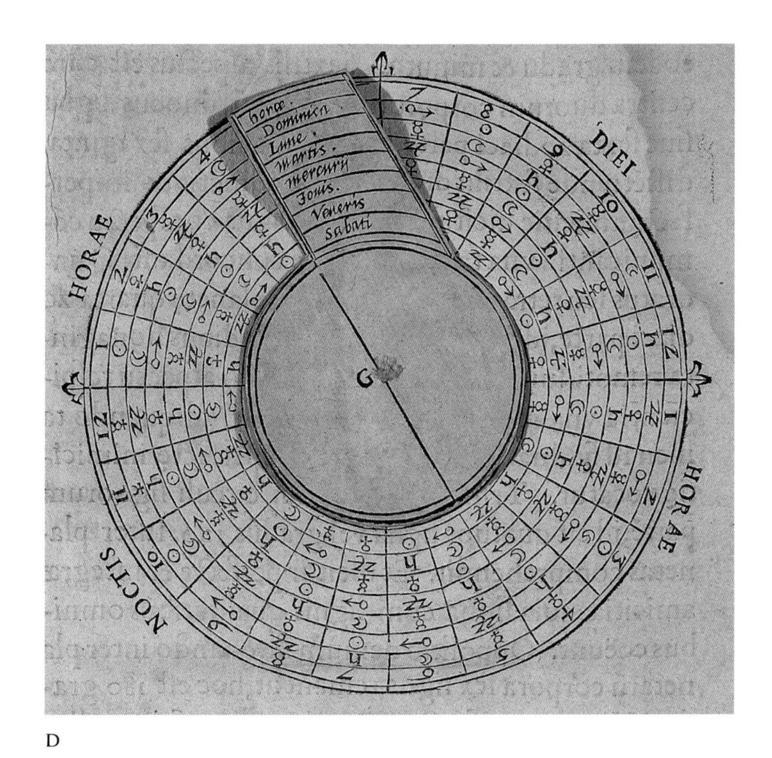

D

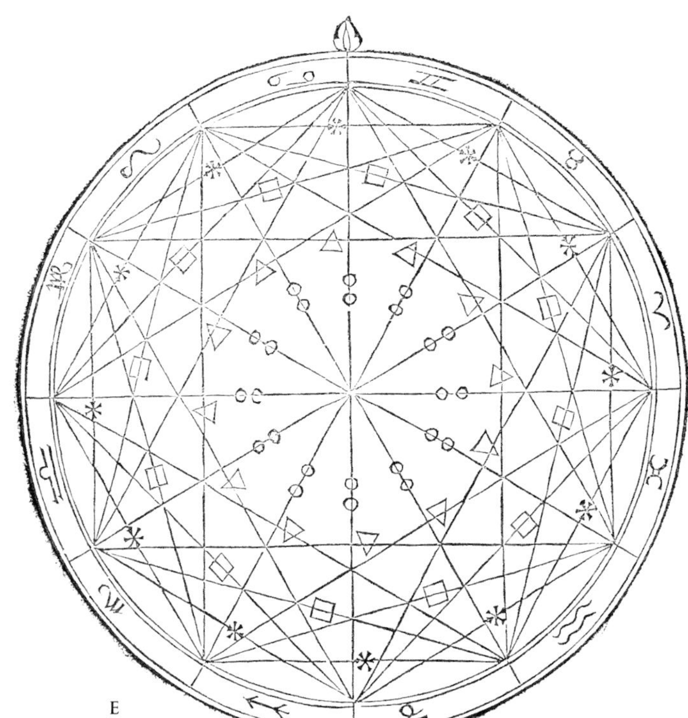

E

houses, at the moment for which a chart was drawn, guided the interpretation of 12 broad themes in the chart: life, death, money, friends, enemies, and the like [B].

Finding the cusps mechanically was an approximate procedure and ignored many subtleties of astrological theory, but it was good enough for most astrologers, most of the time. Handbooks provided all the technical information [C, D, E] that a practitioner needed to draw up a chart or horoscope [F], with supple advice on how to interpret it.

People of all ages and classes wanted to know what to expect about their health, their fortune, their love lives, the weather, politics. It could be treasonous, however, to prepare a chart for one's king, who surely did not want his enemies to believe he was fated to die soon! Despite sporadic opposition from the Church, most people believed in astrology as long as the world around them seemed fundamentally mysterious.

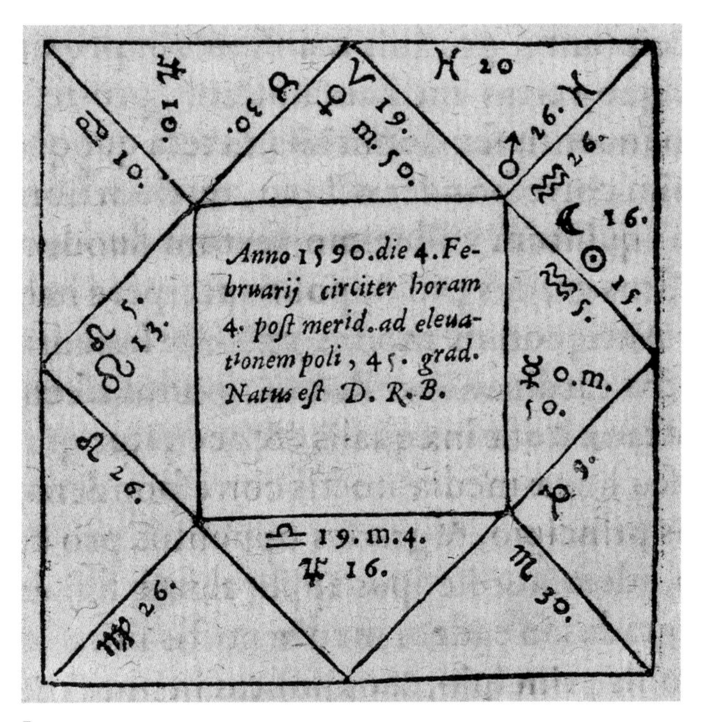

F

CANCER
IUNIUS
IULIUS
LEO
AUGUSTUS
ORBITA SATURNI
SYSTEMA SATURNI
GEMINI
MAIUS
VIRGO
SEPTEMBER
Saturni quinque Satellites ab ejus centro distant diametris annuli Saturnum ambientis.
TAURUS
ORBITA IOVIS
SYSTEMA IOVIALE
LIBRA
OCTOBER
ORBITA MARTIS
ORBITA TERRÆ
Aphelium ♃
Perihelium ♃
ARIES
EX HIS CREATOREM
SCORPIUS
Diametrorum & Magnitudinum Planetariarum respectu Solis et Terræ
PROPORTIONES
calculo Arithmetice deductæ
PISCES
NOVEMBER
SAGITTARIUS
AQUARIUS
FEBRUARIUS
DECEMBER
IANUARIUS
Aphelium ♄
SYSTEMA TYCHONIS

Understanding the Heavens

To understand the heavens is always and everywhere a human goal—to fathom, above all, the dark night sky, populated with stars and a handful of wandering planets. The stars wheel overhead nightly. The Moon waxes and wanes each month. Constellations[+] change with the seasons. Planets follow erratic courses through the zodiac.[+] For most of history, the night sky has served as a kind of laboratory, teasing watchers below into a growing comprehension of their cosmos.[+]

A first step toward understanding the sky was to map its appearance. In the great age of celestial globes and atlases,[+] fanciful renditions of the constellations gave way to elaborate star charts that displayed constellation figures on a background of precisely located stars.

People thought, quite naturally, that our Earth was the center of the starry sphere. In the early 1500s, Nicolaus Copernicus proposed that the Sun was the center of the universe, and the Earth a mere planet. In 1609, Galileo turned the recently invented telescope to the sky, where the new and unexpected things he saw set his mind racing. Many of these novelties lent support to the Copernican theory, but other writers raised doubts, perfectly reasonable doubts, about whether Galileo's instrument had deceived him. Over the course of the century, scientific opinion shifted in favor of the new Sun-centered cosmology,[+] encouraged by Isaac Newton's physics and equally, perhaps, by the changing of the generations!

Meanwhile, the educated public acquired a taste for knowledge of the heavens. Astronomers and natural philosophers responded with books that explained Newton's theories, and craftsmen produced models of the newly understood planetary system. Popular interest in astronomy has continued strong, as ever larger telescopes have extended our vision into the Milky Way[+] galaxy and the wider universe beyond.

The Heavens Mapped

Once astronomers had carefully measured the positions of the stars, they could record them on maps of the heavens. After plotting each star, on either a round celestial globe or a flat star chart, they divided the heavens into regions by drawing the corresponding constellation figures around groups of stars. Globes always represent the entire heavens, yet star charts often fracture the heavens into pieces. Using many different map projections,+ star charts range from those exhibiting the whole sky to atlases showing one map for each constellation. Craftsmen worked partial maps of the heavens into the design and function of certain astronomical instruments, such as astrolabes (see pp. 108–9), quadrants, and nocturnals (see pp. 62–63).

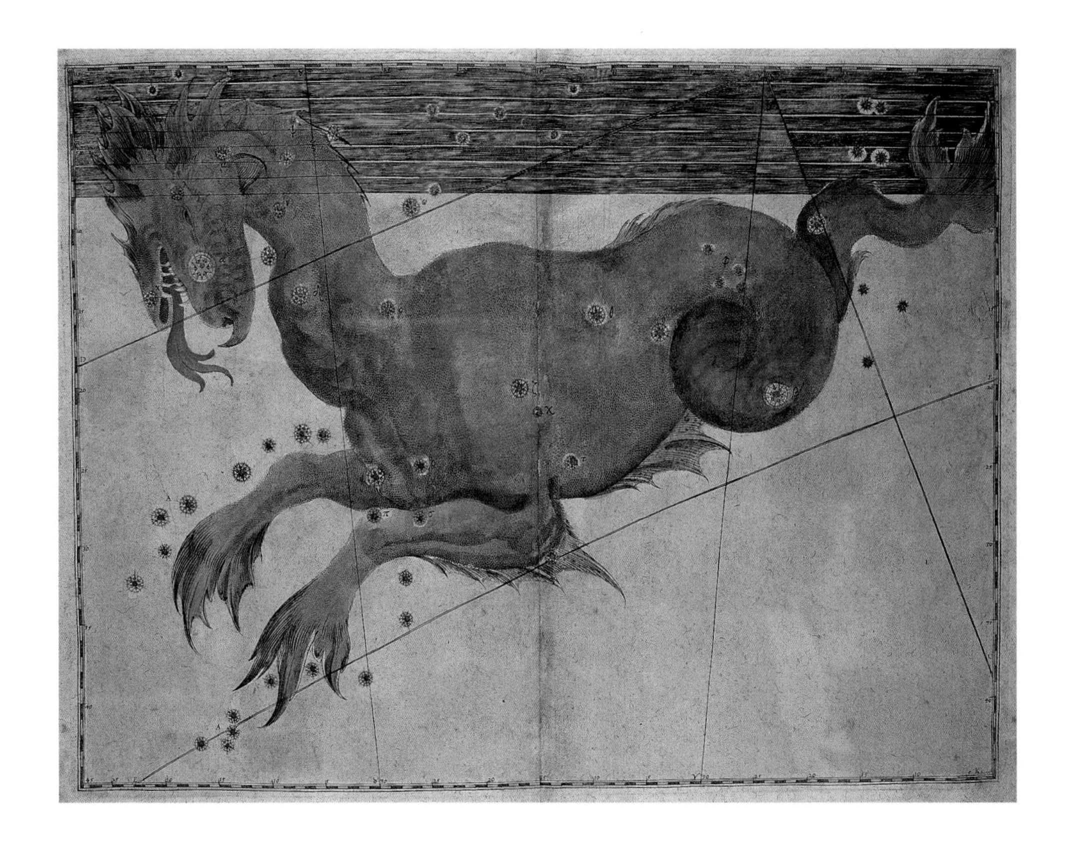

A

Celestial Globes

Celestial globes re-create the starry sphere from an imaginary point in space outside the heavens. Not only a map of the heavens, a globe functions as an instrument. It can demonstrate which constellations are visible at a certain time and location on a particular date, or can determine the time of sunrise and sunset.

People in ancient Greece, the Islamic world [B], China (see p. 26 C), Japan, India, Western Europe [A, D, E], and America have all made globes. The earliest were sculpted marble. Instrument makers crafted later globes from metal with the stars and constellations inscribed directly on the surface [A, B]. But metal globes had their limitations—they were expensive to

B

make and too heavy to be portable, unless a tiny size.

A less expensive, but more complicated way of creating globes evolved in the early 1500s. Hand-colored, printed paper maps of the stars, called gores, were pasted to the surface of a wooden or papier-mâché framework coated with plaster [D, E]. Gore-based globes offered additional benefits—they could be constructed in extremely large sizes and weighed far less than comparably sized metal globes. The maker could also produce virtually identical multiple globes.

Some gores survive today that were never assembled into globes. When laid out flat, the gores create a discontinuous map of the heavens [C].

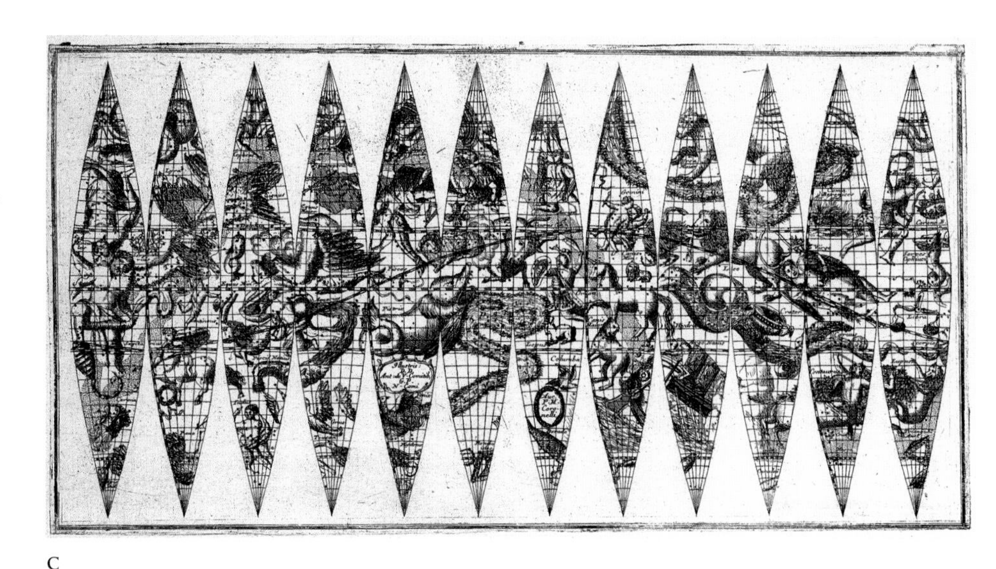

C

D

E

Star Charts

To depict the heavens in their entirety is relatively simple on a globe but considerably more difficult on a flat map. Astronomers developed many map projections[+] to devise easily readable charts of the stars and constellations.[+]

From the first appearance of printed celestial charts in 1515 through the 1700s, the prevalent method among celestial cartographers splits the sky in half using the stereographic projection (see appendix, p. 1451). The heavens divide along the ecliptic[+] into northern and southern hemispheres, contrary to current practice, which separates the heavens along the celestial equator.[+] Each hemisphere appears on its own circular map, with its ecliptic pole[+] at the center and the zodiacal constellations around the edge. Thus the entire heavens are portrayed by means of two maps, commonly called planispheres,[+] sometimes printed together on the same sheet [A].

But how could one depict the entire sky in just one map? Two methods predominate, though both cause substantial distortions in the relative sizes of the constellations.

A

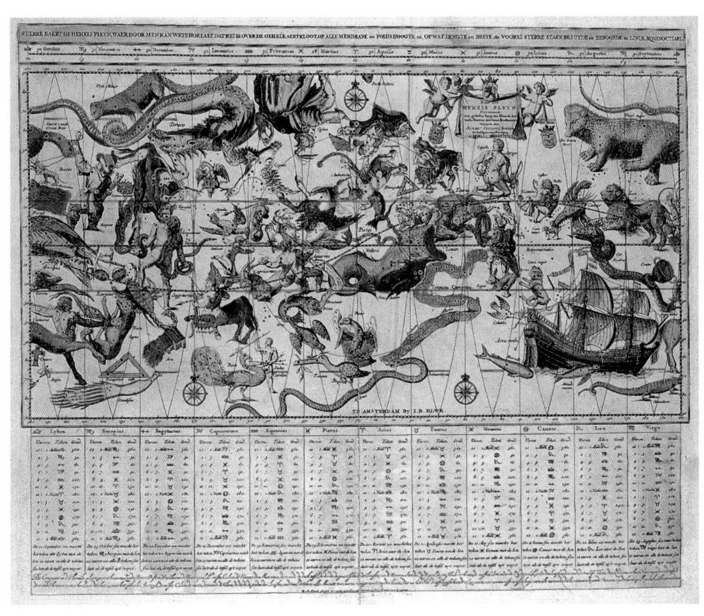

B

C

Some cartographers depict the entire sky in one circular map with the north ecliptic pole at the center [C]. It is as if a hole made at the south ecliptic pole is stretched out to form the outer edge of a circle. This technique places the zodiacal constellations halfway between the edge and the center, with the disproportionately huge southern constellations at the edge.

Other cartographers favor a rectangular format quite similar to the types of maps commonly used to depict the Earth [B]. This style shows the ecliptic and celestial equator⁺ running along the center of the map. The resulting cylindrical projection depicts the zodiacal constellations in proportion, though the northernmost and southernmost constellations appear enormous.

Astrolabe Retes

Astrolabe retes[+] are movable templates of the brightest stars in the night sky. The word "rete" means "net" in Latin, an allusion to the openwork design integral to this part of an astrolabe. The design is functional—an astrolabe user looks beneath the rete (see appendix, p. 142 B) to the gridlines inscribed on a plate, the tympan.[+]

Using the stereographic projection (see appendix, p. 145 I), an astrolabe maker maps the sky from the north celestial pole[+] (at the center) to the Tropic of Capricorn[+] (around the outside edge)—a region containing the brightest stars visible from a typical European or Asian latitude.[+]

One of the many functions of an astrolabe is that of star finder. In fact, its name derives from two Greek words that mean just this: *astro* (star) and *labio* (finder or taker). With the

A

B

C

D

rete positioned for a certain time (see pp. 64–67), the astrolabe displays a corresponding map of the stars. Or the rete can be set to reflect the position of a particular star, yielding the current locations of other stars.

Astrolabe makers exhibited great ingenuity in how their instruments pointed to the stars. Often the pointers are very simple in design [A]. Sometimes, decorative touches such as stars [D] or balls [B] embellish them. Elaborate shapes ornament other star pointers, taking the form of foliage [C], animals, or pointing hands [E]. One example uses serpents' tails as pointers [F]. In another, a dog's head points to three stars—the tongue leading to one and the ears to two others [G]. Occasionally, astrolabe retes map other celestial objects, such as comets [H].

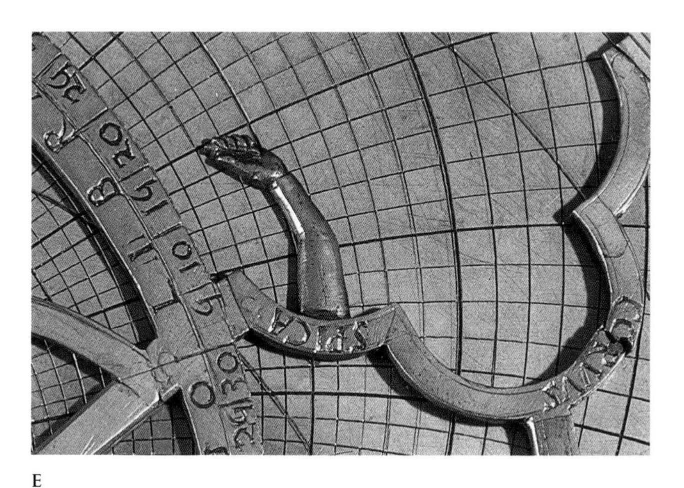

E

F

G

H

A Central Earth

Common sense firmly supported the long-standing belief in the Earth's central position in the cosmos.[+] Anyone could see that the stars revolve overhead and the Earth stands still below. Early European woodcuts and engravings showed Earth, surrounded by the other Aristotelian elements—water, air, and fire—all enclosed by eight or more celestial spheres carrying the Sun, Moon, planets, and stars. Earth did not occupy the place of honor in this tableau. People interpreted the phrase "heaven above" literally, believing that God dwelled high in the sky with the angels and saints, in an outermost sphere beyond those of the planets and stars. The Earth at the center was filthy and corrupt, suited only to be the home of sinners. In this view of the world, the planets and stars circling overhead were pure and eternal, and astronomers truly studied higher things.

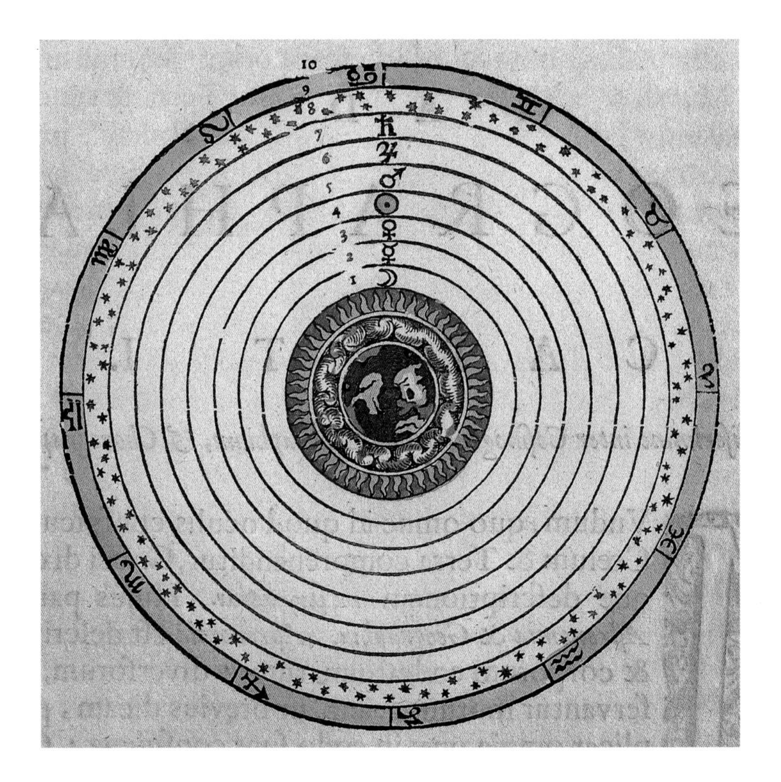

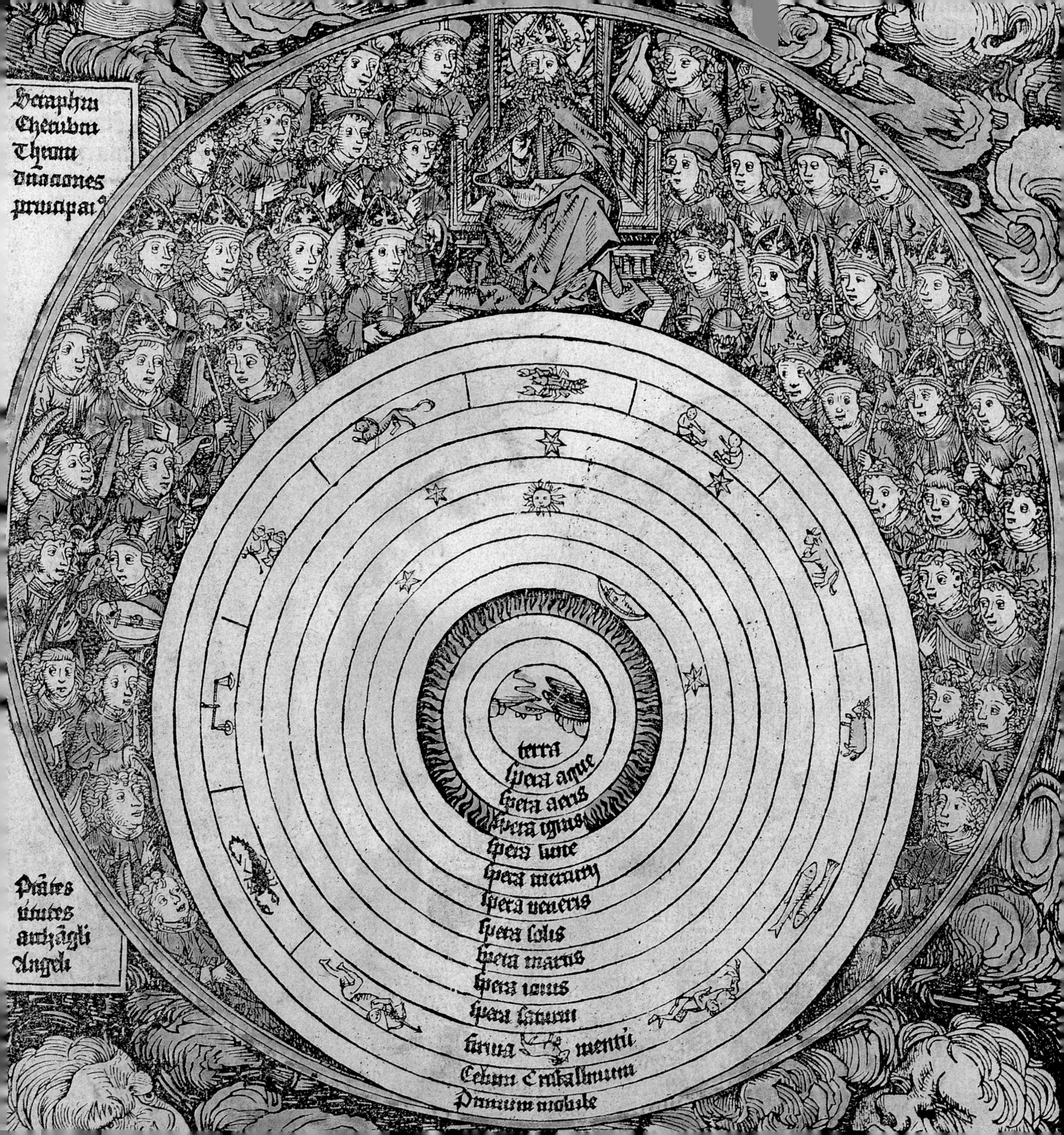

Seraphin
Cherubin
Throni
Dñaciones
Principat9
Ptãtes
Virtutes
Archãgli
Angeli
terra
spera aque
spera aeris
spera ignis
spera lune
spera mercurij
spera veneris
spera solis
spera martis
spera iouis
spera saturni
firma mentu
Celum Cristallinum
Primum mobile

Armillary Spheres

The elegant rings and bands of an armillary sphere [A, B] symbolize the astronomy of a past age. The armillary sphere takes its name from the Latin *armilla,* meaning a bracelet or metal ring. With our Earth firmly at the center, the rings trace out the structure that a patient observer can learn to see in the night sky without the aid of a telescope.

The outer bands, fixed on the stand that supports the device, show the observer's horizon⁺—the horizontal band—and the observer's meridian,⁺ an imaginary circle running overhead from north to south [C]. The horizon ring separates the visible sky from what is hidden beneath the Earth. Inside these unmoving bands is a cagelike assembly of rings. The cage rotates, imitating the sky turning above us. The rotating cage is tilted to show how the heavens rise at an angle from the eastern horizon, swing high in the southern sky, and slant back down into the western horizon (see appendix, p. 142 A).

A

B

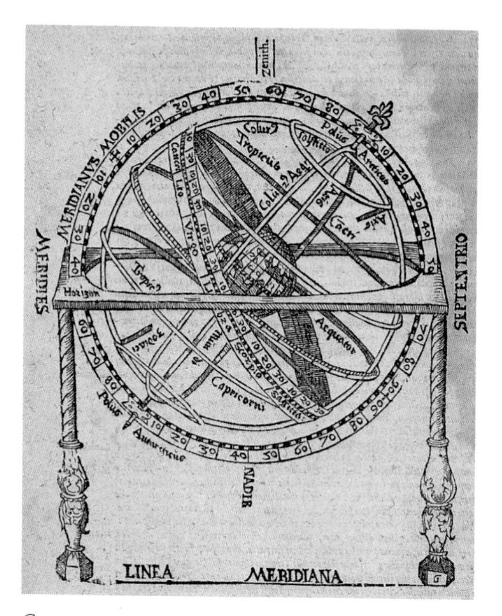

C

D

E

The zodiac[+] is invariably present as a broad band, set at an angle to all the other rings, and marked with its 12 signs. It represents the region in which the Sun and Moon seem to circle us and through which the planets wander. Because the zodiac is angled, some parts of it (especially Gemini and Cancer) cross the horizon ring farther north than others (especially Sagittarius and Capricorn). Seeing how this happens on an armillary sphere is perhaps the easiest way to understand the rising and setting points of the Sun and Moon in different seasons of the year.

An armillary sphere sometimes goes beyond these basic notions to show a tiny Moon, Sun, or planets circling the Earth [D]. Occasionally a craftsman has added gearwork [F], making the instrument a forerunner of later mechanical models of the solar system such as planetaria and orreries (see pp. 134–35).

F

Spheres of the Planets

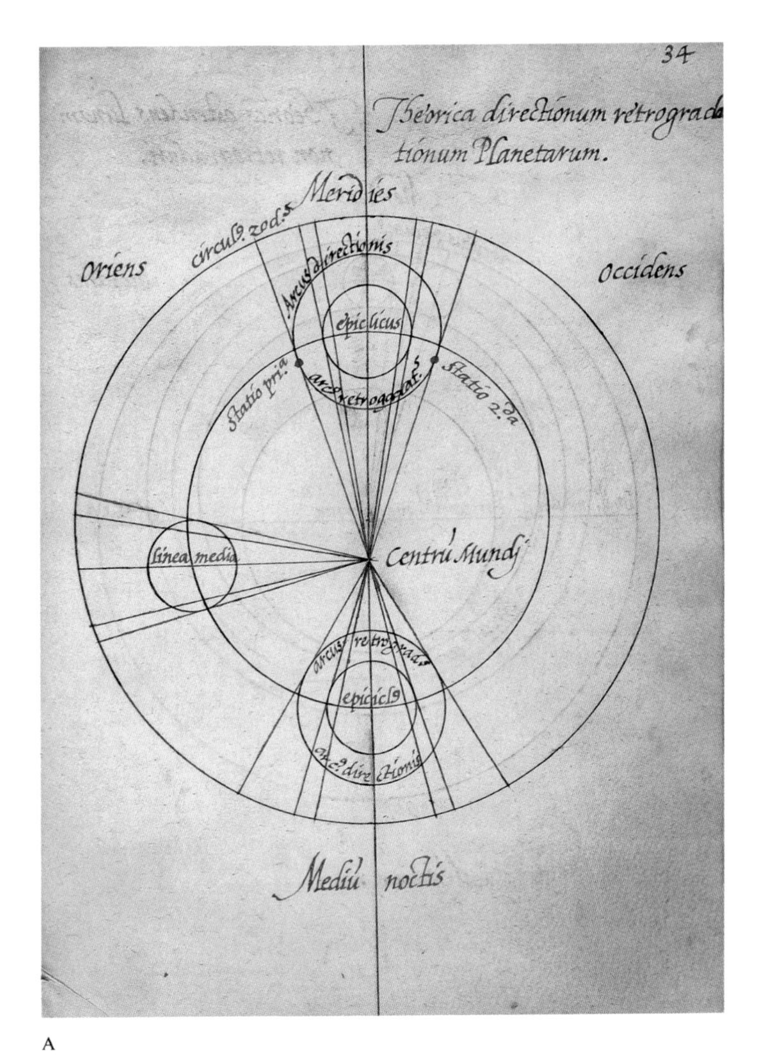

A

The Earth-centered theories of Ptolemy were actually quite accurate, for their time, in their description of the apparent motions of the planets. In his hypotheses, Ptolemy employed off-center circles called eccentrics and epicycles [A]. These worked fine on paper, but did not translate into three-dimensional models showing the actual mechanism that carried a planet around in its intricate Ptolemaic path.

The problem was that circles crossed each other in the geometric models. If one circle rotated, the ones crossing it were sheared in half. With a little ingenuity, however, mathematicians converted two-dimensional geometry into three-dimensional solid geometry. Books explained how these arrangements worked with diagrams depicting the nests of spherical shells [B, C]. A small circle became a ball, and the larger circle carrying it became a spherical shell. Both ball and shell could rotate without interfering with each other.

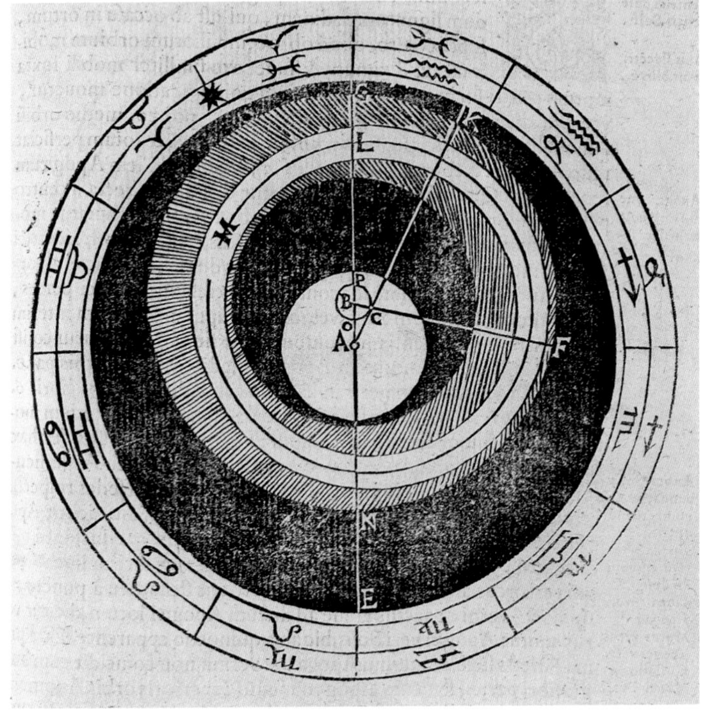

B

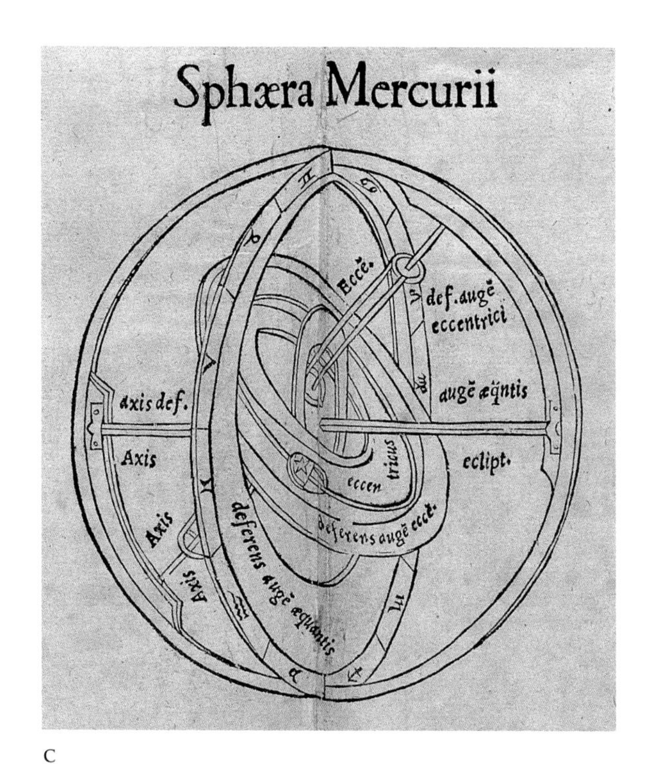

C

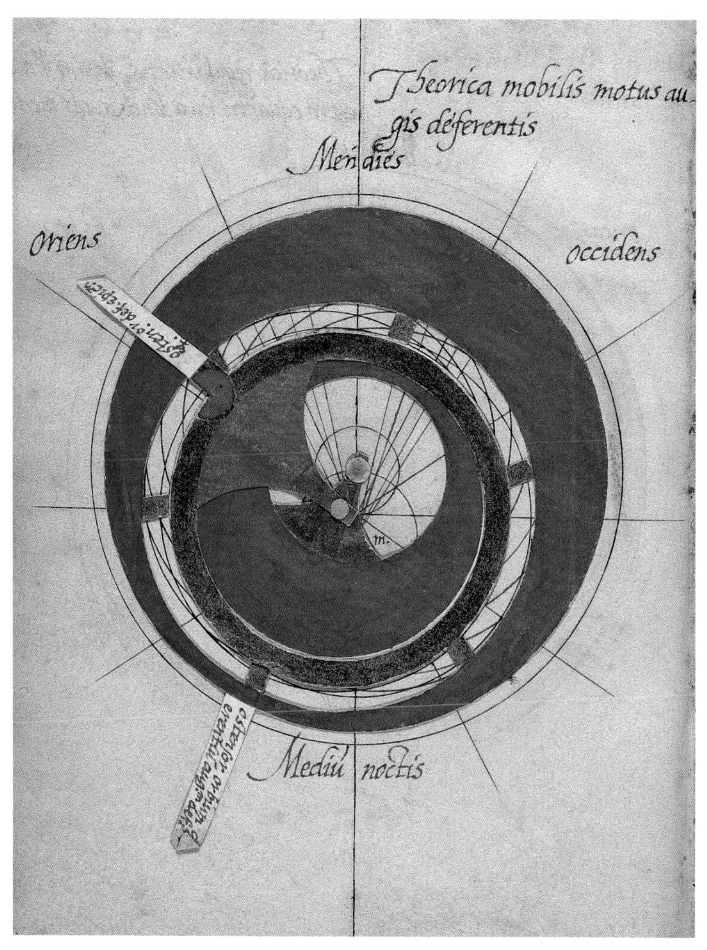

D

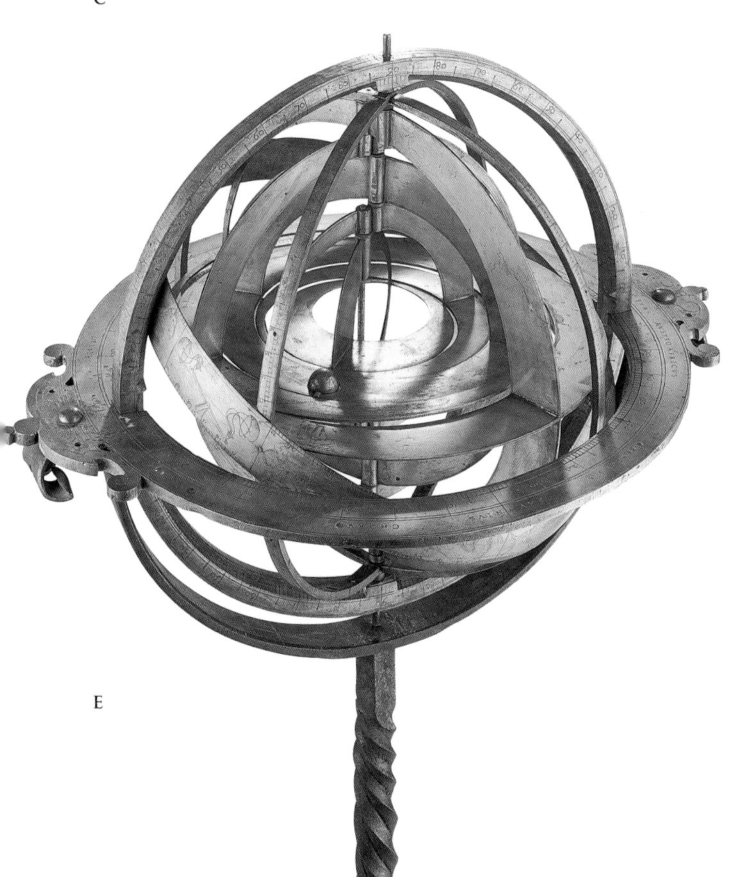
E

To display the geometry more graphically, books sometimes used volvelles,[+] rotating paper models held together with threads [D]. In a sense, books containing volvelles were themselves scientific instruments!

Constructing a model to show the details of a planet's motions, as opposed to the mere fact of its circulation, proved a challenging task. A craftsman could represent the horizontally spinning shells with vertical disks showing their cross sections. This sometimes led to quite impressive mechanisms, as in a sixteenth-century realization of the model for the planet Mercury [E].

Rearranging the Heavens

The quietly revolutionary theories that Nicolaus Copernicus published in 1543, the year of his death, did not immediately convince the world. By rearranging the heavens to put the Sun at the center, and sending the Earth spinning through space among the mere planets, Copernicus sparked enormous long-term changes in our view of the cosmos⁺ and of our place in it. As Copernican ideas took hold in the decades after 1600, the seven planetary models of earlier astronomy merged into a single new concept: the solar system. The difference was huge, for no longer did an unbridgeable chasm separate earthly and celestial knowledge. What we learned on Earth could explain things we saw in the sky, and the heavens could become an exotic laboratory for mundane science.

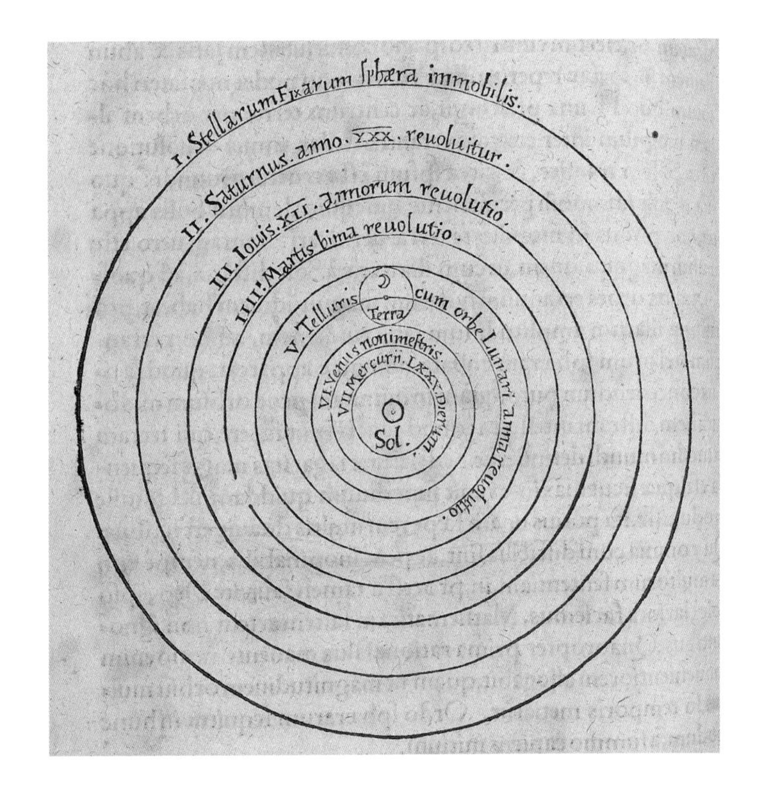

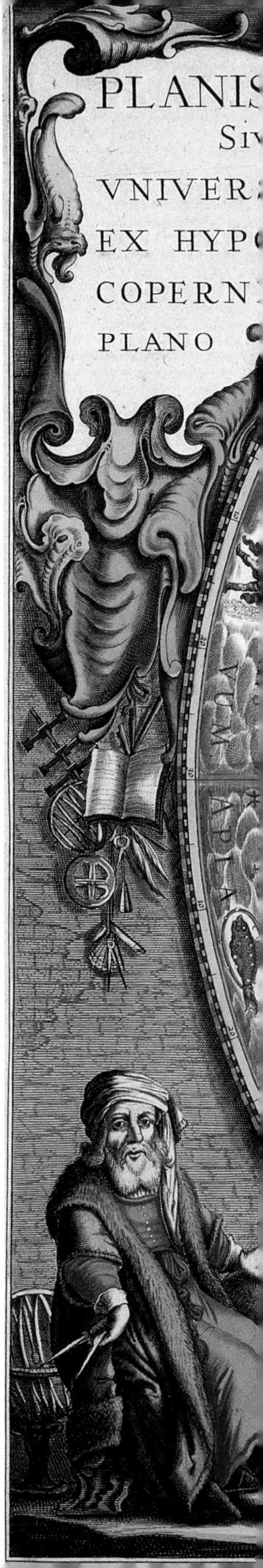

Videbo Cælos tuos,
Sæculum
Ponderibus librata suis
Erigor dum Corrigor
Cl. Ptolem.

Comparison of Systems

By 1600, astronomers (and a growing public audience) had not two but three choices for an overall system of the world. The Earth-centered Ptolemaic cosmos⁺ [A] was familiar and easy to understand, although its inability to explain the coordination among different planetary models had come under increasing scrutiny. The Sun-centered Copernican system [B] united the models for all the planets into a single solar system and offered elegant explanations for many Ptolemaic coincidences, but it insisted on the seemingly absurd idea that the Earth was spinning through space. The new Tychonic compromise [C] avoided the worst problems of the other systems but was ill-proportioned in requiring the entire solar system to revolve around the Earth.

Astronomers addressed these great issues with relish. Johannes Kepler pointed out that a planet's actual motion was surprisingly complicated in the systems of Ptolemy and Tycho. The Copernican system supposed that much of the apparent complexity arose from the motion of our observing platform,

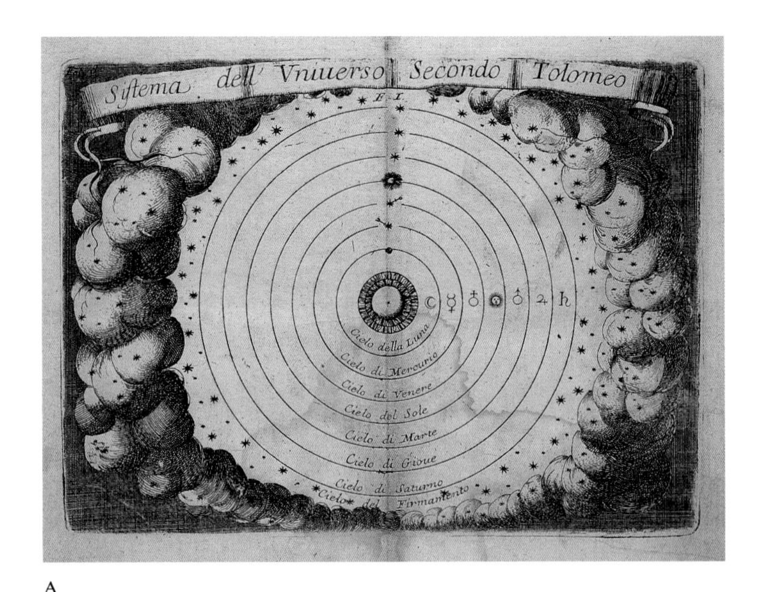

A

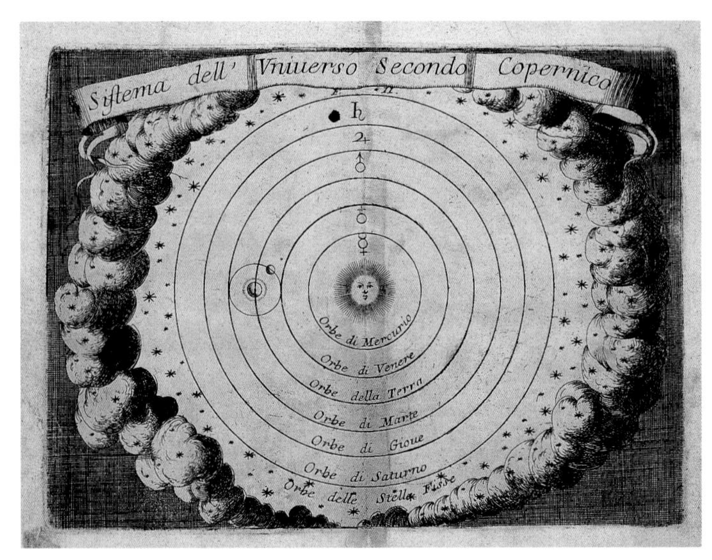

B

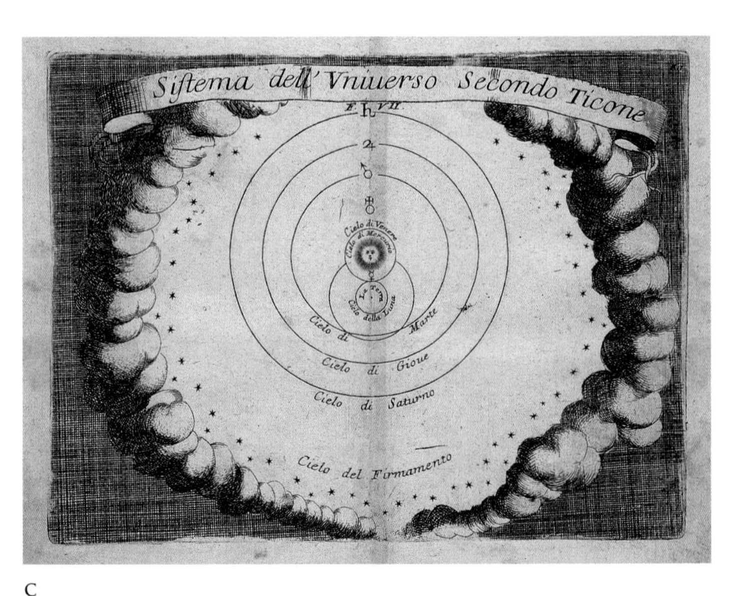

C

D

the Earth. This assumption stripped away the ungainly Ptolemaic epicycles (see pp. 114–15), leaving each planet to travel a single, simplified path through space, its "orbit." This new way of looking at the heavens caught on slowly, because astronomers had always tried simply to explain the sky's appearance, not to understand the physical motion of planets.

Around 1630, René Descartes placed the Sun and its planets within a universe of whirling vortices [D]. His system remained popular into the 1700s, especially in continental Europe.

As new generations learned to study the orbits of the planets, the weight of astronomical opinion shifted in favor of the Sun-centered model. Astronomers drew the looping orbits that the Earth-centered theories implied [E]. The senseless intricacy of these orbits, which pre-Copernican astronomers had never noticed or cared about, gave credence to the followers of Copernicus.

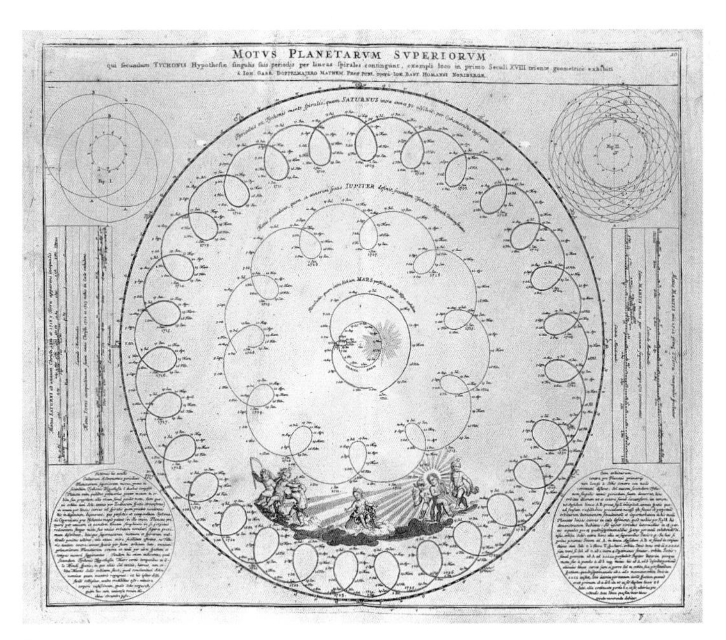

E

Galileo and the Inquisition

A

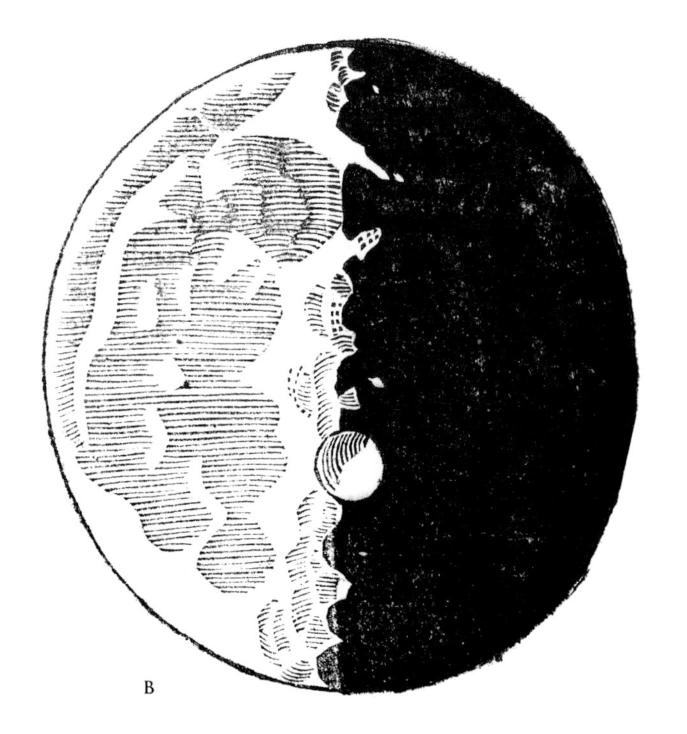

B

In a way, the Catholic Church's condemnation of Galileo for teaching Copernican astronomy came as a surprise to all concerned. Galileo had turned the newly invented telescope [A] to the sky in 1609 and soon discovered the mountainous and cratered surface of the moon [B], the multitudes of stars making up "nebulous" patches [D], the moons of Jupiter [E], and the phases of Venus. He rightly interpreted these phenomena as evidence for the Copernican system, and before long he was disputing the interpretation of scriptural passages that seem to contradict it. In 1616, an ecclesiastical commission instructed him neither to hold nor to defend the doctrine of the Earth's motion.

Seven years later, Maffeo Barberini was elected Pope Urban VIII. Barberini was an urbane man, well-disposed to natural philosophers, and a friend and patron of Galileo. He granted Galileo permission to write on cosmology,⁺ provided that both the Earth-centered and Sun-centered world systems were discussed impartially and hypothetically.

In 1632, Galileo published his *Dialogue . . . concerning the Two Chief World Systems*, in which an intelligent character presented arguments in favor of Copernicus, while a rather stupid Aristotelian philosopher argued against the new ideas.

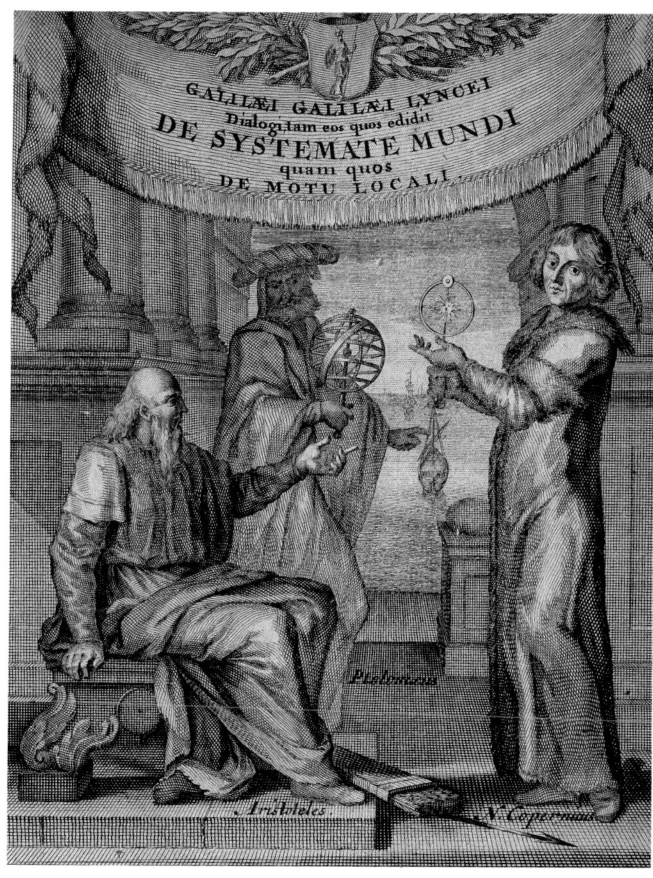

C

D

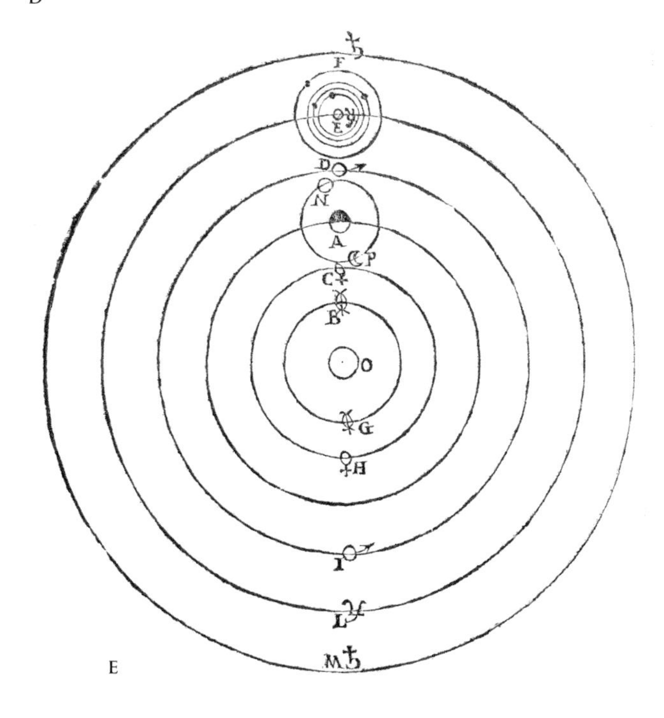

E

The *Dialogue* was a brilliant polemical work, written in Italian to reach a larger audience than would any learned work in Latin, although a Latin edition finally was published in 1699 [C]. It appeared, however, during the bloody religious conflict known as the Thirty Years' War. Pope Urban was already under great pressure for being soft on heretics, and some of Galileo's sarcasm seemed targeted at the pope personally. Enraged, Urban summoned Galileo before the Roman Inquisition, which promptly convicted him and ordered that he be shown the instruments of torture—a lesser sentence than actual torture. Galileo spent the remaining years of his life under house arrest.

Fine-Tuning Order

Copernicus had shown the way to a new understanding of our universe. By putting the Sun at the center, he had revealed an unexpected order in the movements of the planets. If a moving Earth could be accepted, then simple geometry showed why planets sometimes seemed to move backward in the sky. All the planets joined into a single system and lined up in their proper order from the Sun. Yet astronomy was still cluttered with scraps of the older models, for Copernicus had described a Sun-centered world in a Ptolemaic mathematical language. It fell to Johannes Kepler to create an astronomy that was Sun-centered from the inside out, and to Isaac Newton to discover the physical laws by which it all moved. By 1700, they (and others) had established the basic workings of the solar system's celestial clockwork. Astronomers of the next century turned their attention to the tasks of resolving the many details and of spreading the word far and wide.

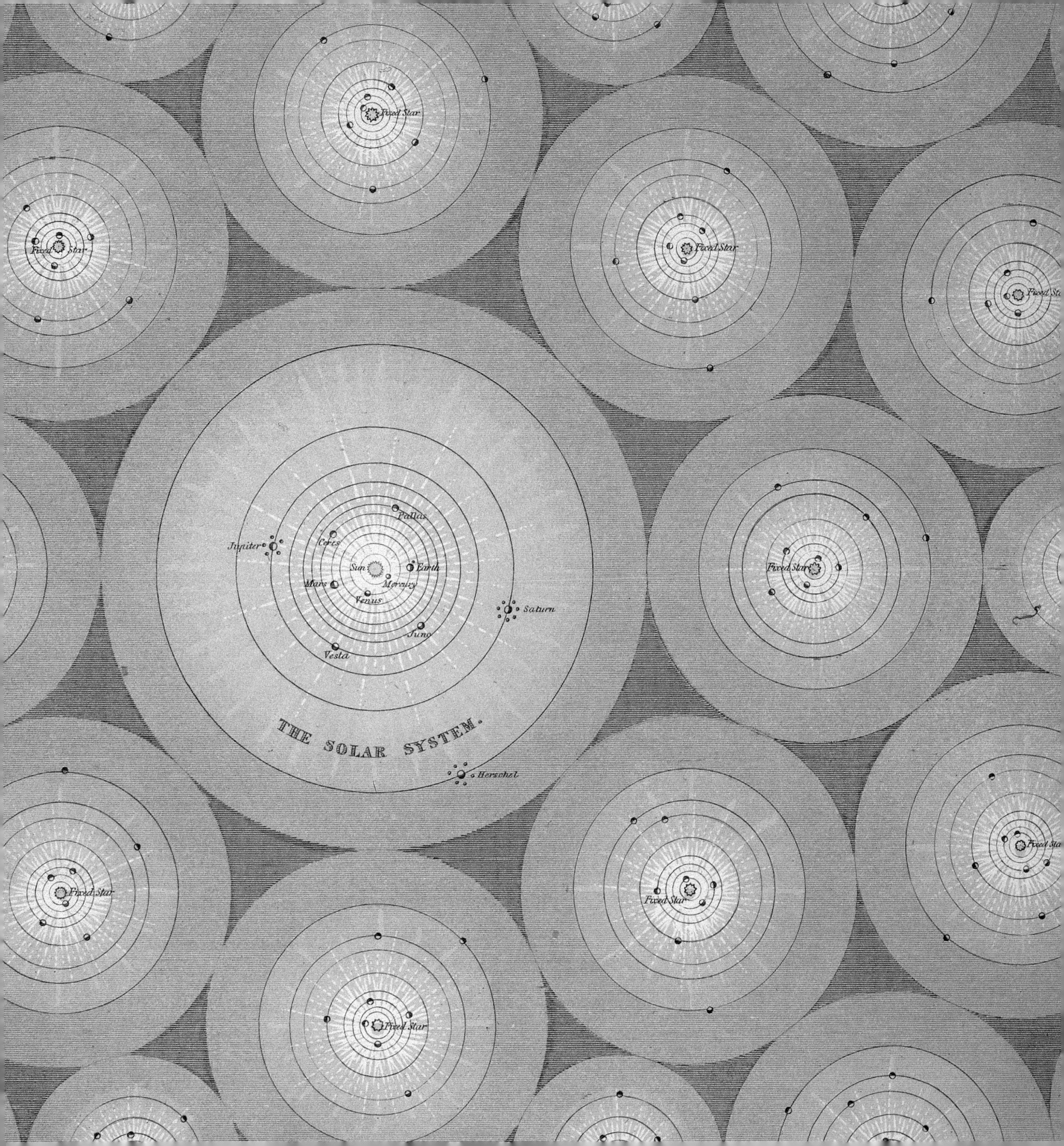
Fixed Star
Fixed Star
Fixed Star
Jupiter
Pallas
Ceres
Sun
Earth
Mars
Mercury
Venus
Saturn
Juno
Vesta
Fixed Star
THE SOLAR SYSTEM.
Herschel
Fixed Star
Fixed Star
Fixed Star

The New Order of the Heavens

Although Copernicus rearranged Ptolemy's theories into a coherent system, his fundamental principles and mathematical techniques remained entirely true to ancient Greek astronomy. In the early 1600s, Johannes Kepler completed the Copernican revolution in astronomy by describing a consistently Sun-centered world in mathematical terms. Near the end of the century, Isaac Newton explained the physical laws that governed the world Kepler had described.

Kepler and Newton were both highly accomplished mathematicians. It was their investigations of the forces governing the motion of the planets, however, that led to a new way of studying the world. Their work marked the birth of physical astronomy.

Kepler began with Copernicus's results but cast aside many of the concepts by which the earlier astronomer had reached them. Instead he concentrated on what he saw as the essential

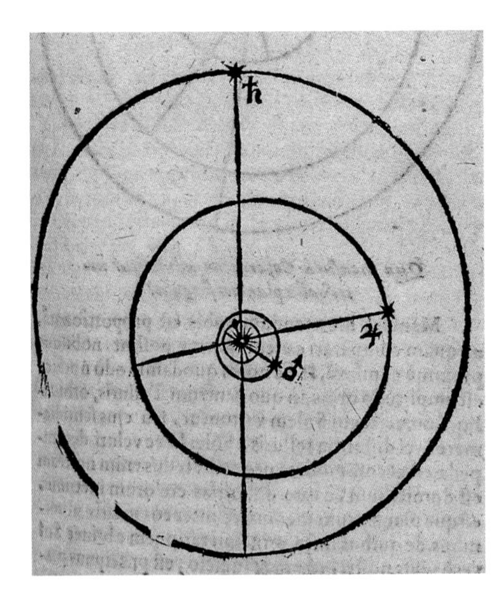

B

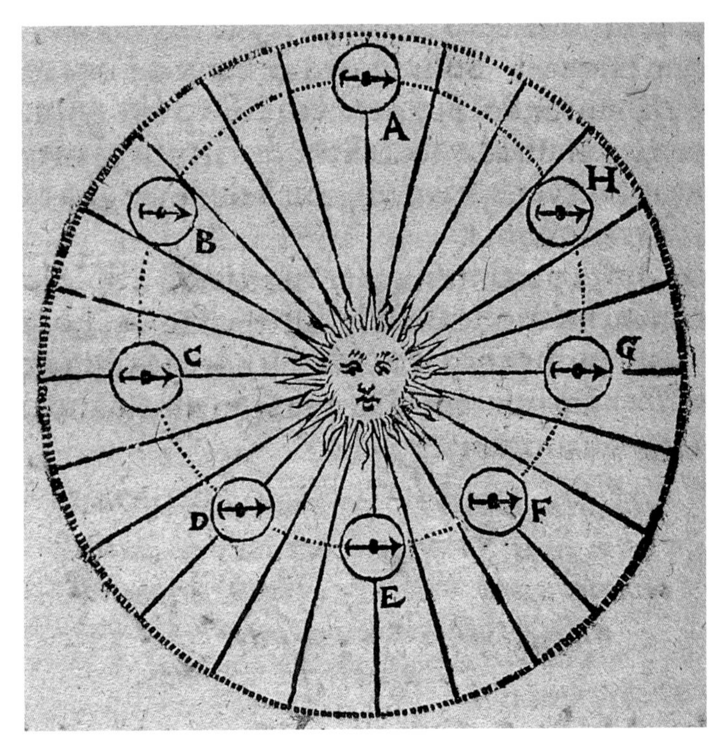

A

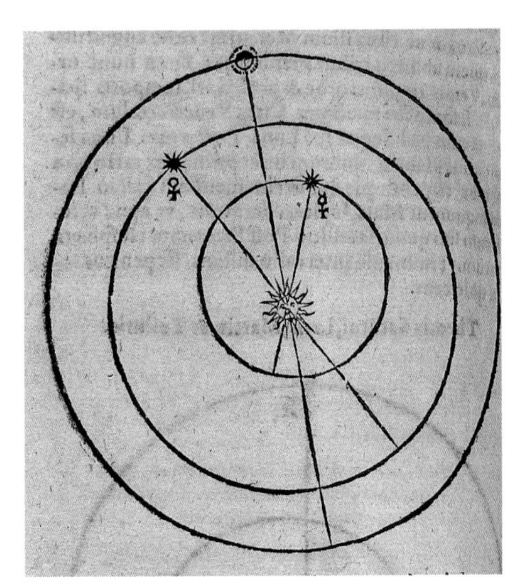

C

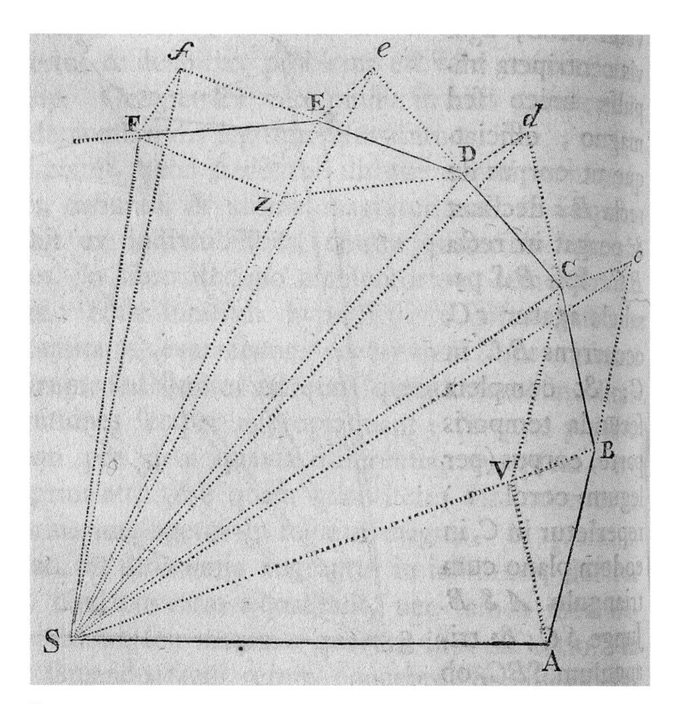

D

phenomena of planetary motion: all the major planets orbited the Sun; planets closer to the Sun moved faster than planets far from it; and each planet moved faster when its own orbit brought it closer to the Sun [B, C]. To Kepler, the implications were clear. A force originating in the Sun moved the planets, and moved them more swiftly when they approached the Sun [A]. By carefully analyzing Tycho's observations, and considering how a solar force might act, Kepler established mathematical rules describing the regular patterns by which the planets move.

Isaac Newton in turn began with Kepler's final results but discarded the concepts on which the earlier theory relied. Newton studied motion in general, building on the work of Galileo and his pupils. His book *Mathematical Principles of Natural Philosophy* [D, E] established a genuinely new order of the heavens and a comprehensive theory of motion within which the movement of heavenly bodies was but an application. For 200 years, theoretical astronomy rested almost entirely on Newton's laws. Even today, most calculations of orbital motion depend upon them.

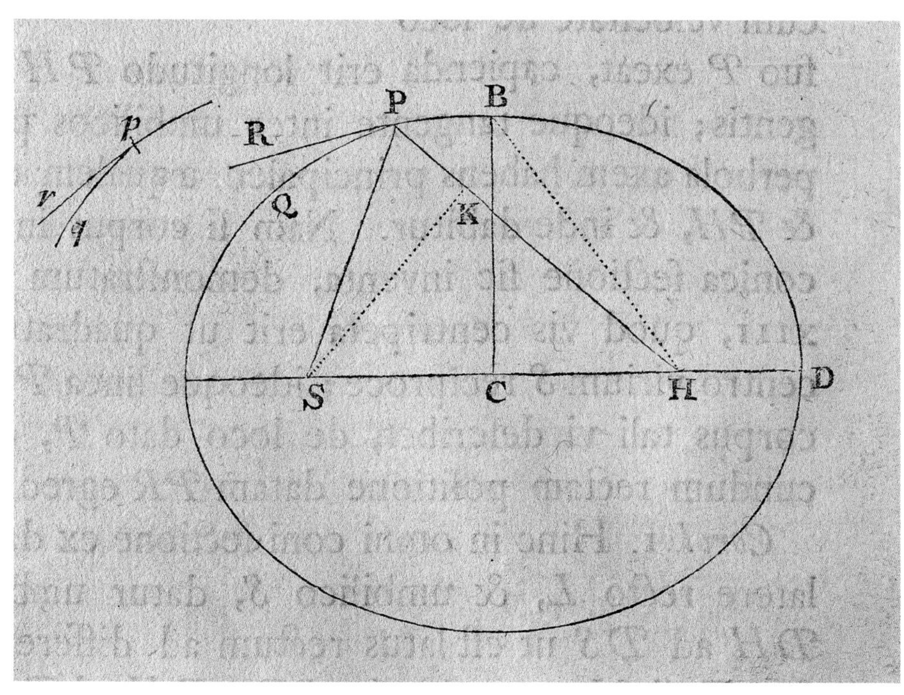

E

Cometary Motion

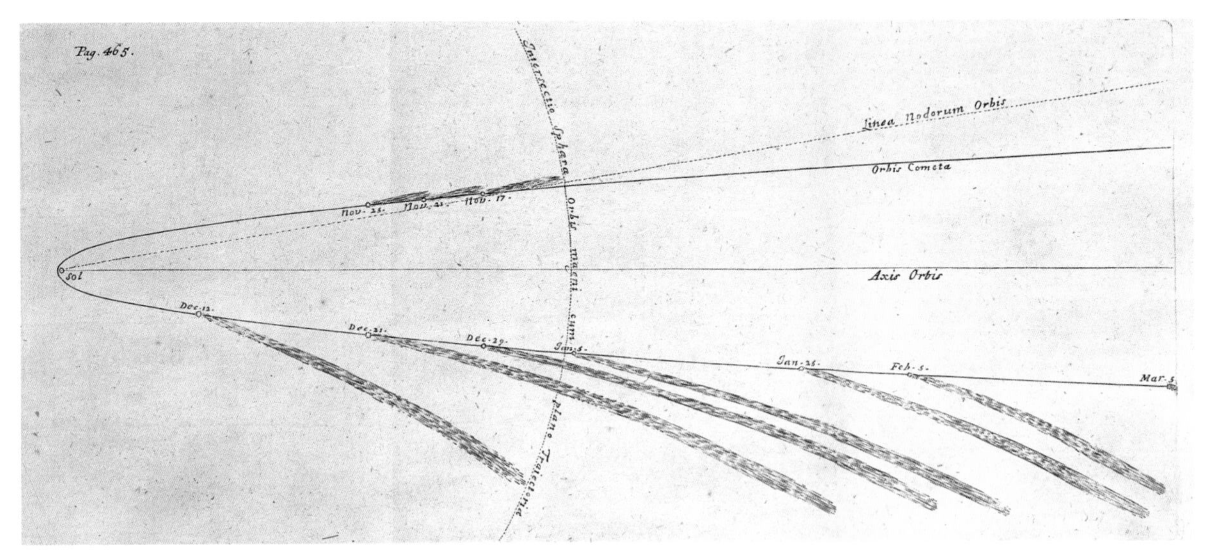

A

B

In 1687, Isaac Newton proved that comets travel around the Sun much like planets, but in more elongated orbits [A]. Edmond Halley subsequently postulated that the eccentric elliptical orbits of certain comets would periodically bring them back near Earth. He correctly predicted the return of the comet of 1682, confirmed 15 years after his death, when an amateur astronomer was the first to observe the comet. Called Halley's Comet, it continues to reappear about every 76 years.

Calculating cometary orbits became a popular astronomical activity during the late 1700s. Astronomers recorded these computations on comet orbit maps [B]. Comets do not necessarily travel around the Sun on the same plane as the Earth, but these maps project their orbits onto it. Orbit maps often tell about the period⁺ of the comet, if applicable.

Instrument makers constructed cometaria [C] to demonstrate cometary orbits. A crank on the side of such a machine turns an arm in a circular pattern around a central Sun.

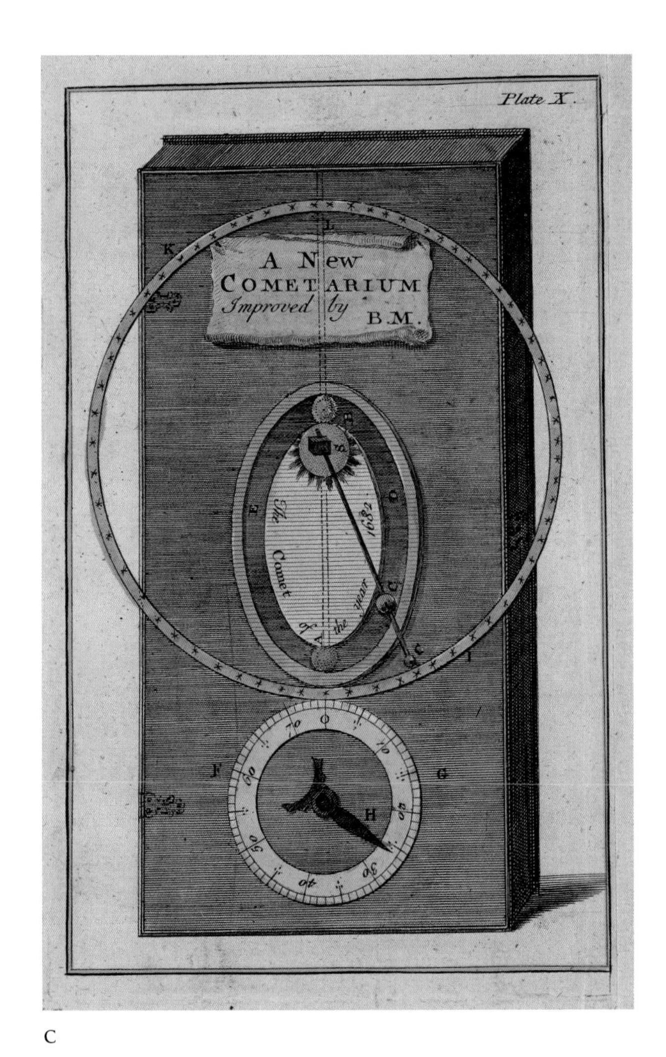

C

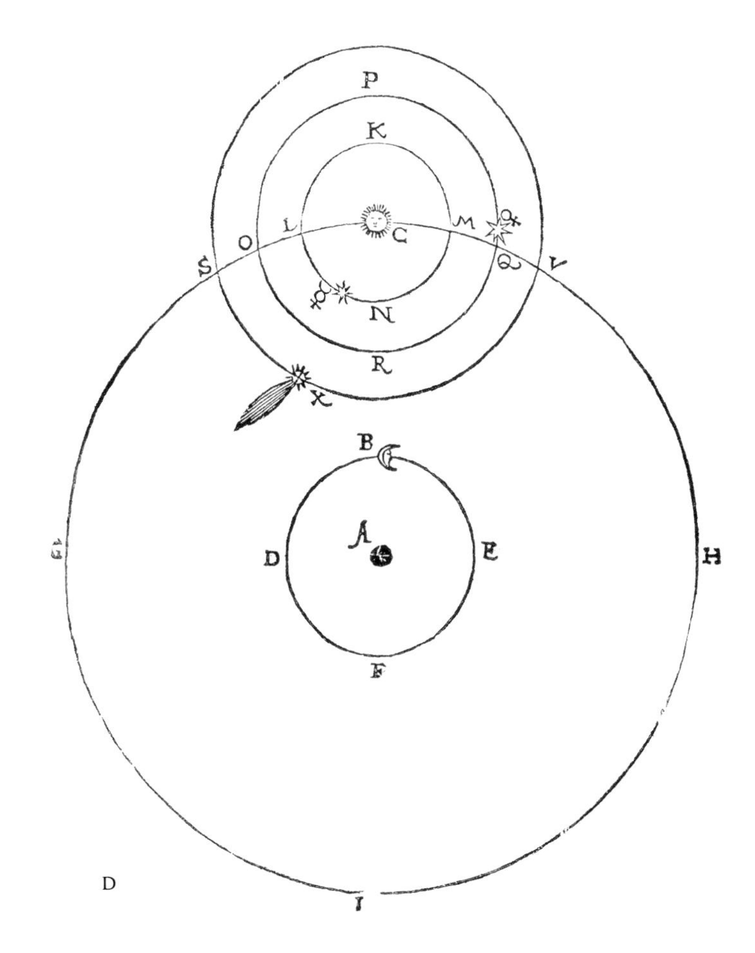

D

Attached to this arm is an orb, representing a comet, which is set in an elliptical groove around the central sun. As the crank is turned, the comet moves in its elliptical path, traveling the length of the arm as it orbits.

Originally, people categorized comets as meteorological phenomena, not as celestial bodies. Aristotle postulated that comets resided in the spheres of fire and air surrounding the Earth and just below the Moon [E]. This theory of cometary nature and location lasted until 1577, when Tycho Brahe proved that comets move in the celestial realm [D].

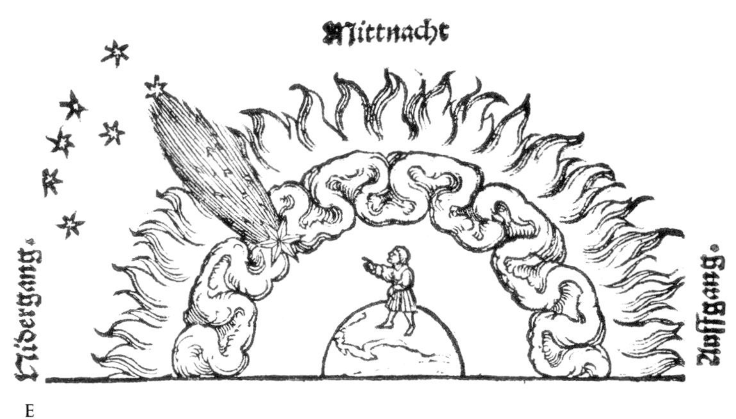

E

Mechanized Heavens

The mechanized planetarium, one of the most popular scientific objects of the 1700s, especially in Britain, displayed the motion of the planets around the Sun. The operation of the device's carefully crafted mechanisms inspired awe and wonder, and a sense of the universe's divinely imposed stable order. Although the cost of planetaria usually restricted their distribution to wealthy patrons, scientifically minded entrepreneurs might still advertise such instruments in the frontispieces to their popular books on astronomy or even raffle off such treasures to those purchasing tickets to their lecture series.

Early Planetaria

Around 1705, George Graham built the first tabletop planetarium [A]. It is, technically, a tellurium, showing the daily, monthly, and annual motions of the Earth and Moon as they orbit the Sun. John Rowley soon made a copy of Graham's apparatus for Charles Boyle, the fourth earl of Orrery, whose title became commonly used as a label for any English mechanical planetarium (see p. 46 A). Depending on its focus and level of detail, an instrument of this type can bear one of several names: planetarium (showing the relative motions of the planets orbiting the Sun), orrery (displaying a tilted Earth-Moon system in orbit around the Sun), grand orrery (including the motions of the planets and their moons around the Sun), tellurium or tellurian (emphasizing the Earth-Sun system) [B], or lunarium (focusing on the Earth-Moon interaction).

B

In Britain, particularly in the middle to late 1700s, orreries contributed to the popularity of astronomy. Astronomical knowledge was a mark of education and social status, and ownership of an orrery gave material evidence of such status. James Ferguson, a well-known instrument maker and the leading popularizer of astronomy in that century, may even have constructed a beautiful orrery for King George III [C].

A

A flurry of discoveries beginning in 1781 added to the popularity of astronomy and to the features included on orreries. William Herschel found a

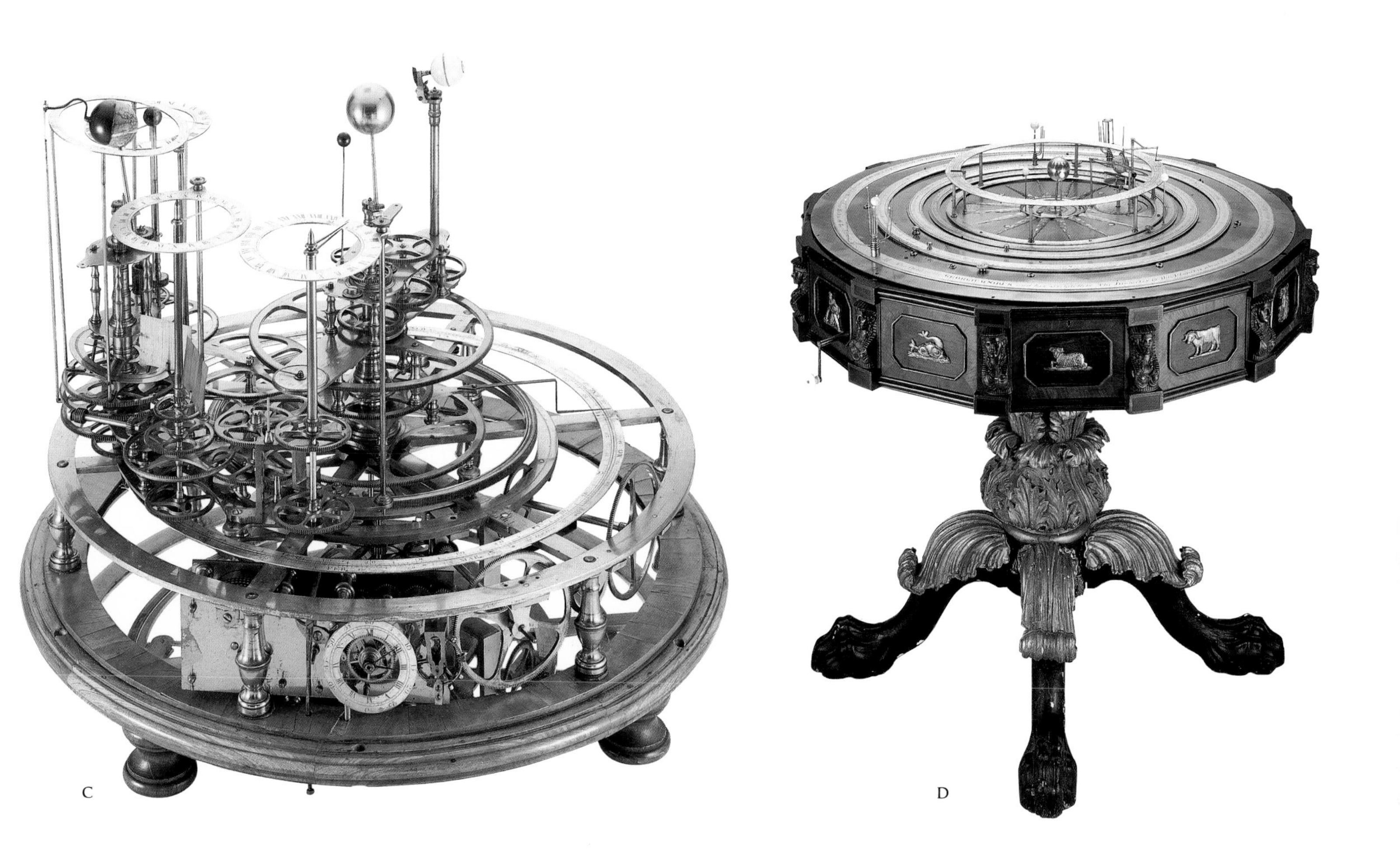

C

D

new planet, Uranus, in 1781, two of its moons in 1787, and a pair of new moons orbiting Saturn two years later. Instrument makers rushed to incorporate these discoveries on their orreries. Thomas Heath's grand orrery [D], made originally in the 1740s, was modified to include Uranus, labeled here as "Georgium Sidus" (see p. 48). It includes not only the two moons that Herschel did discover but also four more he thought he saw [E].

E

The Wider Universe

Over time, telescopes, orreries, and books displayed a universe filled with more and more objects at greater and greater distances. The discovery of Uranus doubled the dimensions of the solar system, which grew another 50 percent with Neptune's sighting in 1846. For centuries, astronomers had concentrated on the motions of the Sun, Moon, and planets. Around 1800, though, the stars themselves became objects of investigation, raising new questions. What is the size and shape of the Milky Way?[+] Are all stars located within the Milky Way? And what are those fuzzy patches, the nebulae?[+] Are they stars, or are they vast collections of stars, like the Milky Way? For more than a century, astronomers struggled to answer these tantalizing questions.

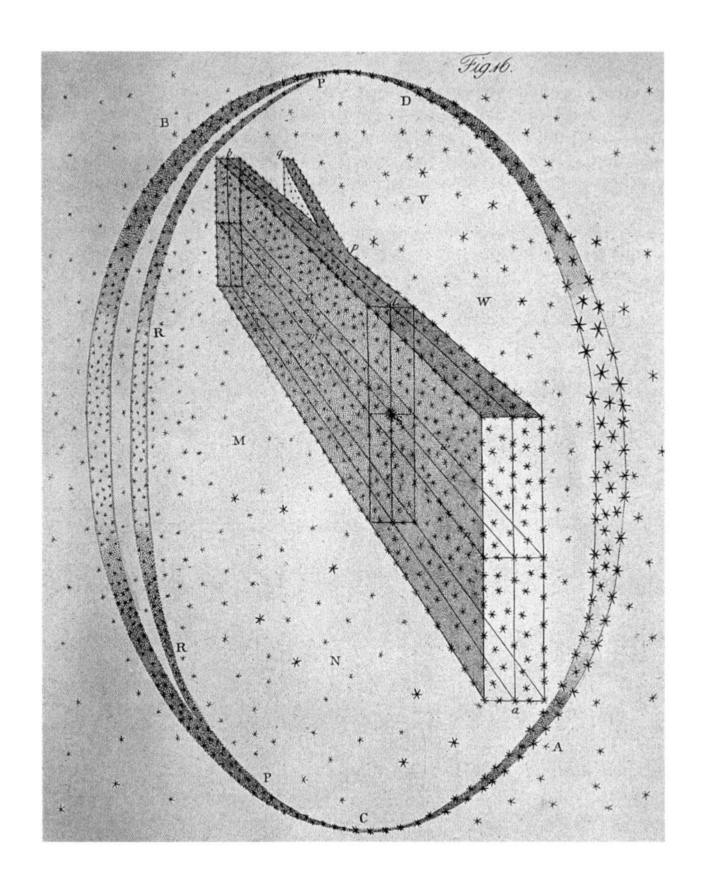

Telescopes and Galaxies

The earliest telescopes were refractors, with lenses placed at both ends of tubes made usually from pasteboard covered with leather [A] or sharkskin. Because the lenses distort images, astronomers in the 1700s preferred reflecting telescopes, which use metal mirrors instead of lenses to gather light (see appendix, p. 143 D).

By the mid-1700s, telescopes were mass-produced, with production num-

A

B

C

D

bers sometimes engraved on them. More ornate telescopes [B], rarely if ever used to observe the heavens, served instead to enhance the prestige of their owners. Lenses and mirrors kept increasing in size [C], and by the end of the century, telescopes revealed additional stars and more information about their motions and distributions.

In the late 1700s, French astronomer Charles Messier catalogued about 100 fuzzy objects he wanted to recognize—and ignore!—as he searched for new comets. In the next century, astronomers used ever larger telescopes to resolve more and more details of the nebulae[+] in Messier's catalogue, including one known simply as M51 [D].

Beginning in the 1860s, astronomers complemented telescopes with photographic plates and spectroscopes, leading to the eventual recording of the positions, movements, and compositions of thousands of stars. By analyzing starlight, astronomers could categorize stars and even distinguish between stars and glowing clouds of gas [E]. Spectroscopy[+] eventually showed that some nebulous patches are indeed galaxies, outside and unimaginably distant from the Milky Way.[+] Astronomers concluded that the universe contains more than 100 billion galaxies, each of which, on average, holds 100 billion stars.

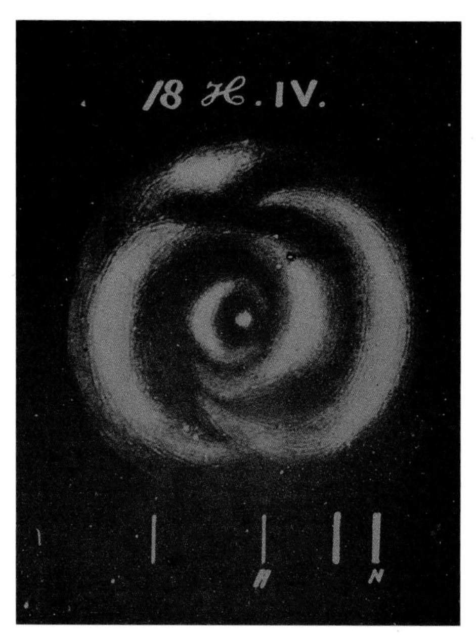

E

Glossary

ALIDADE a rule[+] with sights, attached to an instrument such as an astrolabe (see appendix, p. 142 B) or a surveyor's compass and used to determine the direction of an object

ALMANAC an annual publication containing dates of celestial phenomena (new moons, eclipses, equinoxes, etc.) and sometimes weather forecasts and other predictions

ALTITUDE the angle of elevation of an object in the sky

ASCENDANT the point in the ecliptic[+] that is rising from the eastern horizon at any given moment

ASTERISM a recognizable grouping of stars; part of a constellation[+]

ATLAS a collection of maps, usually published in book form

AZIMUTH a horizontal direction, typically measured as an angle clockwise from north

BROADSIDE a printed sheet of paper bearing information for the public

CELESTIAL SPHERE an immense, imaginary sphere, centered on Earth, and regarded as carrying the stars on its surface

CLOCK TIME uniformly flowing time as measured by an ideal clock (compare with *solar time*[+])

COMPENDIUM (PL. COMPENDIA) a portable combination of several different instruments, typically including one or more sundials, a compass, and sometimes a nocturnal or volvelles.[+]

CONJUNCTION, PLANETARY the apparent grouping of two or more planets at the same celestial longitude,[+] as seen from Earth

CONSTELLATION a group of stars that forms a pattern, often named after the shape or figure it seems to resemble

COSMOGRAPHY the science of describing and mapping the universe

COSMOLOGY the science of the structure and origins of the universe

COSMOS an ancient Greek word meaning the totality of everything that exists

CULMINATION the highest point in the passage of an object across the sky

DECLINATION angular distance north or south of the celestial equator,[+] on the celestial sphere[+]

ECLIPTIC the yearly path of the Sun's apparent motion against the background of stars; the plane in which the Earth and the Sun orbit one another

EQUATOR, CELESTIAL the projection of the terrestrial equator[+] onto the celestial sphere[+]

EQUATOR, TERRESTRIAL a circle around the body of the Earth, midway between the north and south poles

ESCAPEMENT the part of a pendulum clock that regulates the rate of the falling weight and therefore the speed of the clock

GNOMON the part of a sundial (see appendix, p. 143 C) that casts a shadow, indicating the time

HORIZON an idealized circle, centered on the observer, marking the farthest extent of the visible surface of the Earth and defining the horizontal plane

HOROSCOPE in astrology, the point of the ecliptic that is rising from the eastern horizon,[+] or more generally, the configuration of the planets at a specific time

LATITUDE, CELESTIAL on a celestial sphere,[+] angular distance north or south of the ecliptic[+] (compare with *declination*[+])

LATITUDE, TERRESTRIAL angular distance north or south of the terrestrial equator,[+] measured from 0 at the equator to 90 at either pole

LONGITUDE, CELESTIAL angular distance east or west of the vernal equinox, measured in the plane of the ecliptic (compare with *right ascension*[+])

LONGITUDE, TERRESTRIAL angular distance east or west of a standard or "prime" meridian[+]

MAGNITUDE a scale (historically, from one to six) indicating the apparent brightness of a star as seen from Earth; the brighter the star, the smaller the magnitude

MERIDIAN an arc on the Earth or on the celestial sphere[+] extending from the north pole to the south pole, passing through or over the observer's position

MERIDIAN, STANDARD a particular meridian chosen as a reference point; does not necessarily pass through the observer's position

MILKY WAY a faint, patchy white band running across the sky, revealed by telescopes to be an immense, disk-shaped collection of stars

NEBULA (PL. NEBULAE) historically, a fuzzy patch in the sky; used today as a label for immense, glowing clouds of interstellar dust and gas

NOMENCLATURE a system of naming things

NOVA (PL. NOVAE) a new star appearing in the heavens

PARHELION (PL. PARHELIA) bright spots appearing on either side of the sun, sometimes accompanied by halos giving the illusion of multiple suns

PERIOD the number of years it takes a celestial body to complete its orbit

PLANISPHERE a flattened map or pair of maps of the celestial sphere,[+] typically constructed using the stereographic projection (see appendix, p. 145 I)

POLE STAR a reference star located near the celestial pole;[+] the actual star nearest the pole changes over periods of thousands of years, due to precession[+]

POLES, CELESTIAL the projection of the Earth's north and south poles onto the celestial sphere[+]

POLES, ECLIPTIC the points in the sky in the direction perpendicular to the plane of the ecliptic[+]

PRECESSION the slowly changing direction of the Earth's rotational axis, resulting in a gradual drifting of celestial coordinates

PROJECTION, MAP any of a number of ways of representing the spherical Earth or the celestial sphere[+] on a flat map

RETE the rotating, cutaway star map on the face of an astrolabe (see appendix, p. 142 B)

RIGHT ASCENSION the angular distance east or west of a standard meridian,[+] measured in the plane of the celestial equator and often expressed in hours, where 24 hours equals the 360 degrees of a complete revolution

RULE a straight piece of metal or other material that rotates around the central axis of an instrument; used to line up points with the center

SEASONAL HOURS hours defined as one-twelfth of the day or night whose length thus differs between day and night and varies with the changing of the seasons; also called unequal hours

SOLAR TIME time as measured by the daily passage of the Sun

SPECTROSCOPY the analysis of light, with the aid of a device such as a prism or diffraction grating, revealing the material composition of the source of light

STAR CATALOGUE a list of stars with their coordinates, either latitude[+] and longitude,[+] or declination[+] and right ascension[+]

SURVEYING the charting of terrestrial regions, based on distance and angle measurements made with instruments

TROPIC OF CANCER; OF CAPRICORN the circles on a terrestrial globe marking the farthest north and south latitudes of the Sun, which it attains at the solstices; also, the projection of these circles onto the celestial sphere[+]

TYMPAN a plate on the face of an astrolabe (see appendix, p. 142 B) fixed beneath the rotating rete,[+] inscribed with altitude[+] and azimuth[+] coordinates for observers at a particular terrestrial latitude

UNEQUAL HOURS. SEE SEASONAL HOURS

VOLVELLE an instrument composed of rotating disks, in one or more layers, used to demonstrate astronomical phenomena and aid in calculation; often made of paper and assembled in a book or manuscript

ZODIAC the band running around the celestial sphere[+] within which the Sun, Moon, and planets appear to move

Astronomical Appendix

Drawings by Roy A. Kaelin Jr., modified for publication by Vivian Larkins

A. Armillary Sphere

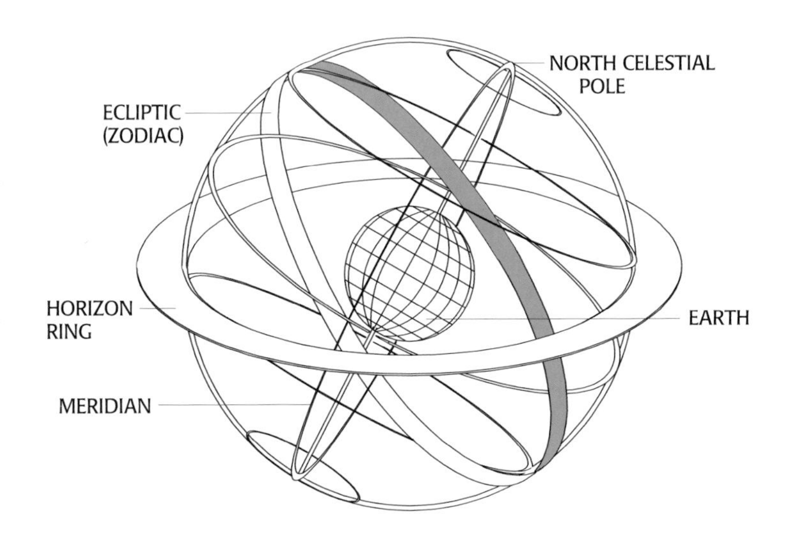

B. Astrolabe

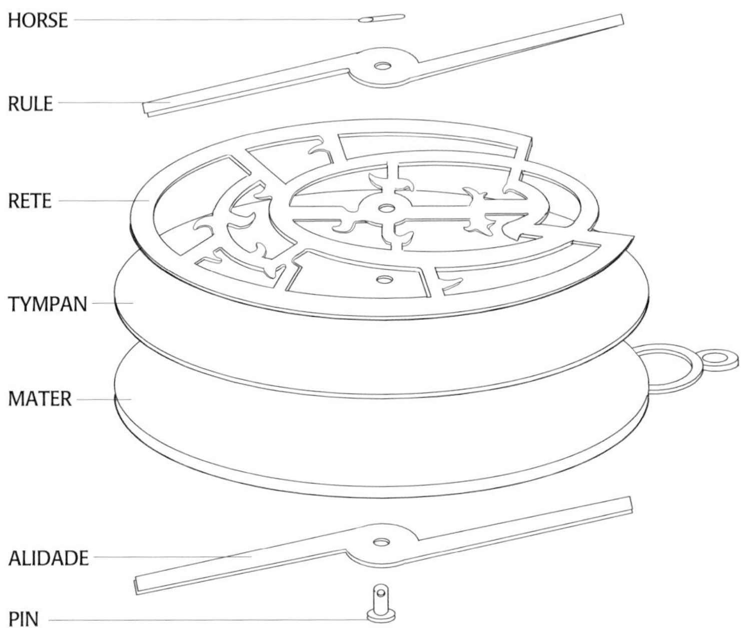

C. Sundial

GNOMON

HOUR SCALE

LATITUDE SCALE

COMPASS

PLUMB BOB

D. Telescopes

REFRACTOR

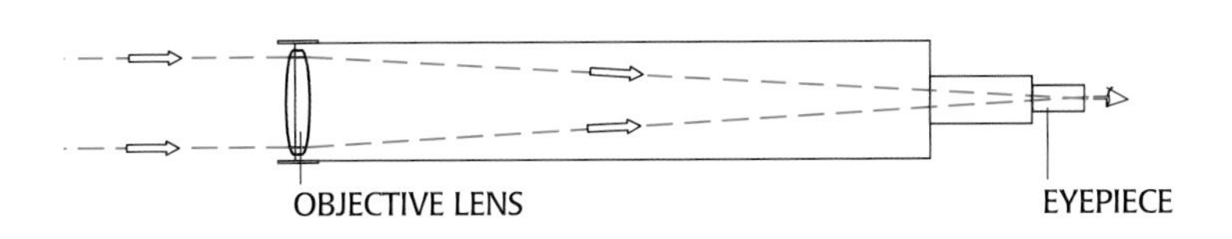

REFLECTOR

EYEPIECE

PRIMARY MIRROR

SECONDARY MIRROR (WITH SUPPORT)

E. Apparent Daily Motion of the Sun on Different Dates

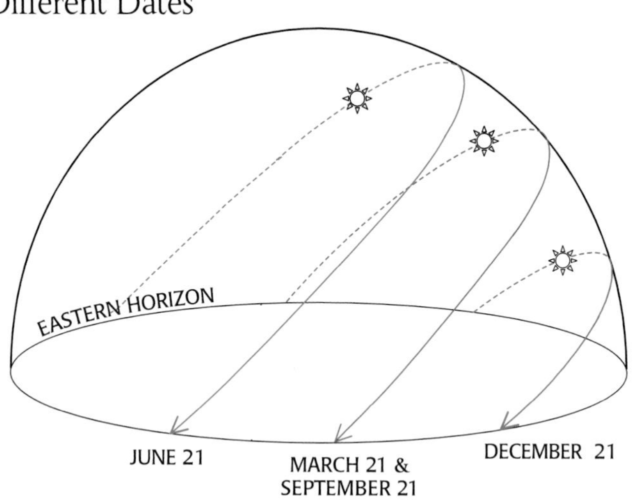

F. Equation of Time

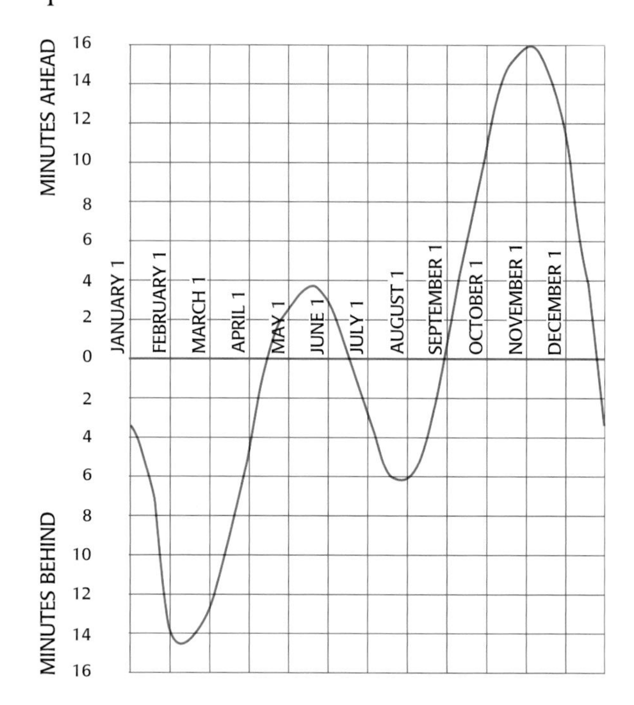

This chart represents the amount of time by which a sundial runs ahead of or behind clock time. Values are approximate.

G. Apparent Motion of the Stars

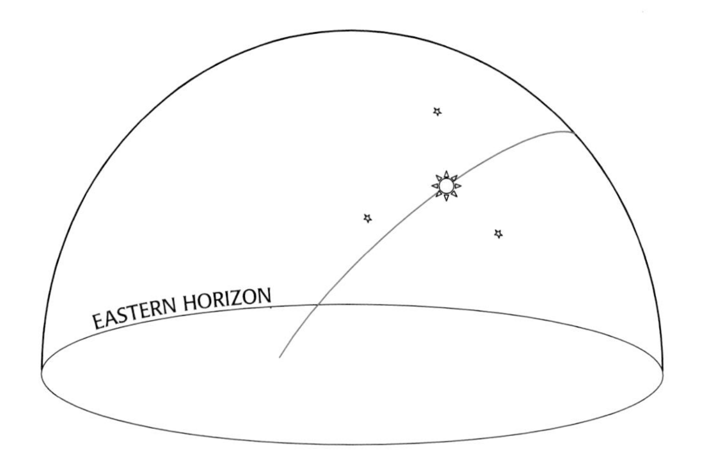

When the constellation rises, it is not visible because the Sun has already risen.

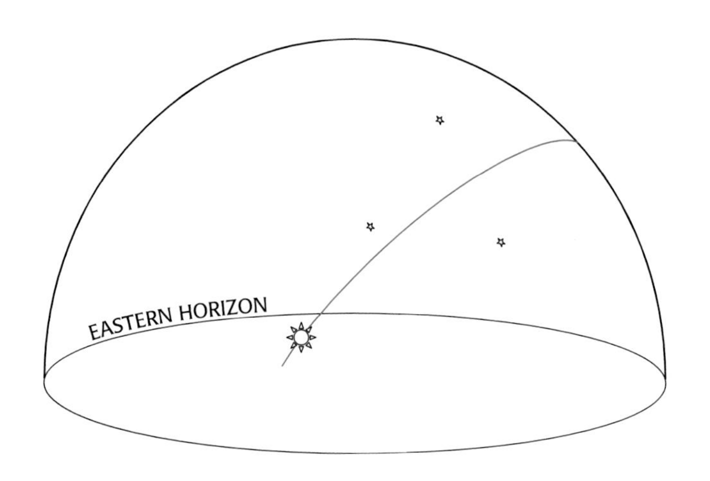

Several days later, the Sun has moved farther east on its yearly path. Therefore when the constellation rises, it *is* visible because the Sun has *not* yet risen.

H. Celestial Latitude and Longitude

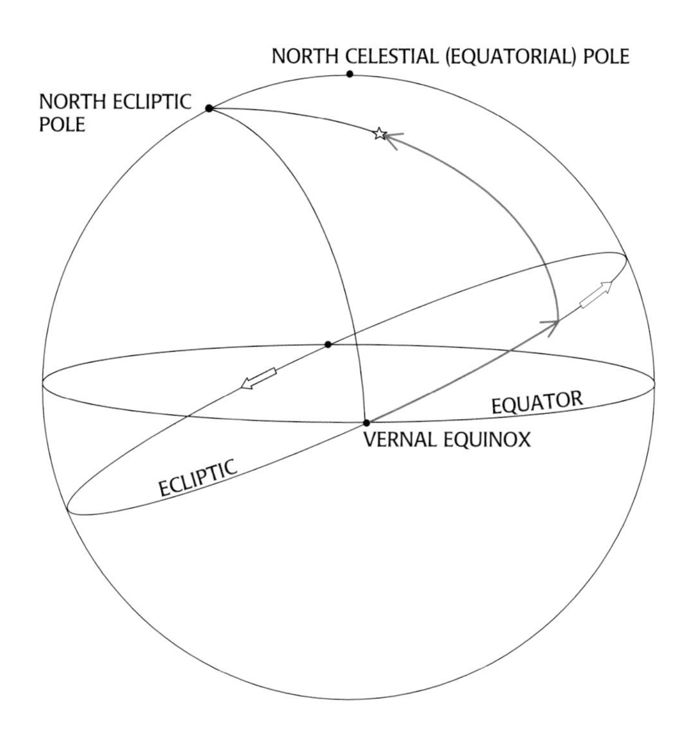

The red arrow indicates celestial longitude; the blue arrow indicates celestial latitude.

I. Stereographic Projection

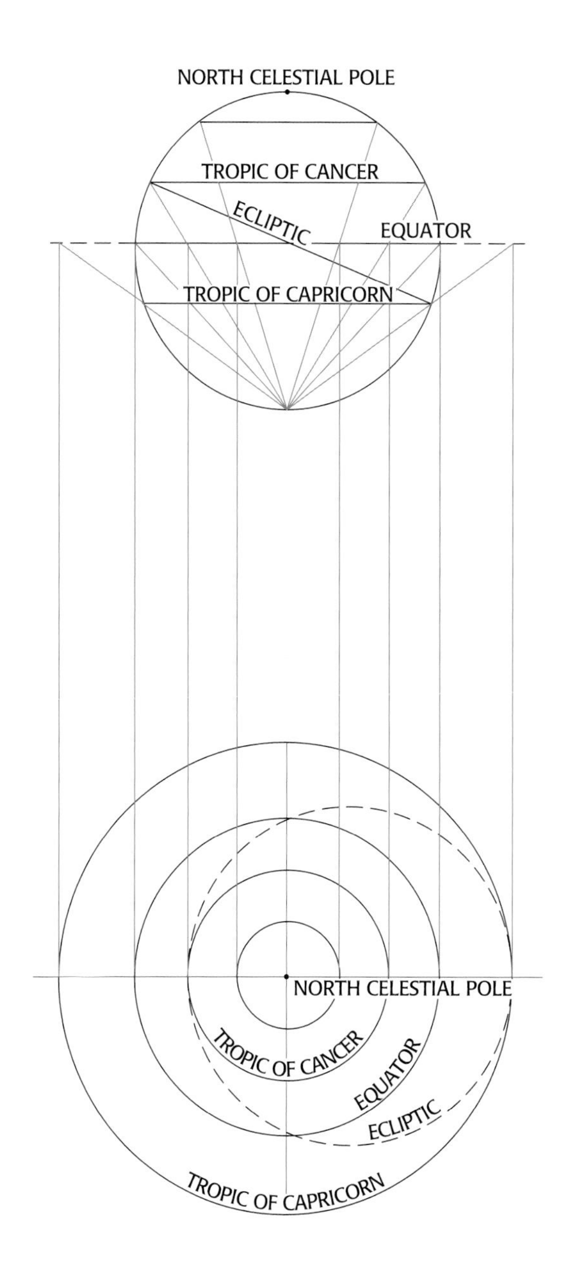

Note: For many star charts, the stereographic projection is centered on the ecliptic pole.

Further Reading

Aveni, Anthony. *Stairway to the Stars: Skywatching in Three Great Ancient Cultures.* New York: John Wiley and Sons, 1997.

Bakich, Michael E. *Cambridge Guide to the Constellations.* New York: Cambridge University Press, 1995.

Bud, Robert, and Deborah Jean Warner, eds. *Instruments of Science: An Historical Encyclopedia.* New York and London: The Science Museum, London, and The National Museum of American History, Smithsonian Institution, in association with Garland Publishing, 1998.

Campion, Nicholas. *An Introduction to the History of Astrology.* London: Institute for the Study of Cycles in World Affairs, 1982.

Cohen, I. Bernard. *Album of Science: From Leonardo to Lavoisier 1450–1800.* New York: Charles Scribner's Sons, 1980.

Crowe, Michael J. *Theories of the World from Antiquity to the Copernican Revolution.* New York: Dover Publications, 1990.

Dekker, Elly. *Globes at Greenwich: A Catalogue of the Globes and Armillary Spheres in the National Maritime Museum.* Oxford: Oxford University Press and the National Maritime Museum, 1999.

Dekker, Elly, and Peter van der Krogt. *Globes from the Western World.* London: Zwemmer, 1993.

Dohrn-van Rossum, Gerhard. *The History of the Hour: Clocks and Modern Temporal Orders.* Chicago: The University of Chicago Press, 1996.

Evans, James. *History and Practice of Ancient Astronomy.* New York: Oxford University Press, 1998.

Ford, Brian J. *Images of Science: A History of Scientific Illustrations.* New York: Oxford University Press, 1993.

Gingerich, Owen. *The Great Copernicus Chase and Other Adventures in Astronomical History.* Cambridge: Cambridge University Press, 1992.

Harley, J. B., and David Woodward, eds. *The History of Cartography.* 2 vols. to date. Chicago: The University of Chicago Press, 1987–.

Heiniger, S. K., Jr. *The Cosmographical Glass: Renaissance Diagrams of the Universe.* San Marino, Calif.: The Huntington Library, 1977.

Hoskin, Michael A., ed. *Cambridge Illustrated History of Astronomy.* New York: Cambridge University Press, 1997.

Hunter, Andrew, ed. *Thornton and Tully's Scientific Books, Libraries, and Collectors.* Aldershot, Hampshire, U.K.: Ashgate, 2000.

King, David. *Astronomy in the Service of Islam.* Aldershot, Hampshire, U.K.: Variorum, 1993.

King, Henry C. *Geared to the Stars: The Evolution of Planetariums, Orreries, and Astronomical Clocks.* Toronto: University of Toronto Press, 1978.

———. *The History of the Telescope.* New York: Dover Publications, 1979.

Krupp, E. C. *Beyond the Blue Horizon: Myths & Legends of the Sun, Moon, Stars, & Planets.* Oxford: Oxford University Press, 1991.

Lankford, John, ed. *History of Astronomy: An Encyclopedia.* New York: Garland Publishing, 1997.

Murdoch, John E. *Album of Science: Antiquity and the Middle Ages.* New York: Charles Scribner's Sons, 1984.

North, John D. *The Norton History of Astronomy and Cosmology.* New York: Norton, 1995. (Published as *The Fontana History of Astronomy and Cosmology* in Europe.)

Pingree, David, et al. *Eastern Astrolabes*, vol. 2 of *Historic Scientific Instruments of the Adler Planetarium.* Chicago: Adler Planetarium & Astronomy Museum, forthcoming (2001).

Schechner, Sara. *Comets: Popular Culture and the Birth of Modern Cosmology.* Princeton, N.J.: Princeton University Press, 1997.

Selin, Helaine, ed. *Encyclopedia of the History of Science, Technology, and Medicine in Non-Western Cultures.* Boston: Kluwer Academic Press, 1997.

Shirley, Rodney W. *The Mapping of the World: Early Printed World Maps 1472–1700.* London: The Holland Press, 1987.

Sobel, Dava, and William J. H. Andrewes. *The Illustrated Longitude.* New York: Walker, 1998.

Turner, Anthony J. *Mathematical Instruments in Antiquity and the Middle Ages.* London: Vade-Mecum Press, 1994.

Turner, Gerard L'E. *Nineteenth-Century Scientific Instruments.* Berkeley: University of California Press, 1998.

———. *Scientific Instruments 1500–1900: An Introduction.* Berkeley: University of California Press, 1991.

Walker, Christopher, ed. *Astronomy before the Telescope.* New York: St. Martin's Press, 1996.

Warner, Deborah J. *The Sky Explored: Celestial Cartography 1500–1800.* New York: Alan R. Liss, 1979.

Waugh, Albert. *Sundials: Their Theory and Construction.* New York: Dover Publications, 1973.

Webster, Roderick, and Marjorie Webster. *Western Astrolabes,* vol. 1 of *Historic Scientific Instruments of the Adler Planetarium.* Chicago: Adler Planetarium & Astronomy Museum, 1998.

Yeomans, Donald K. *Comets: A Chronological History of Observation, Science, Myth, and Folklore.* New York: John Wiley & Sons 1991.

Web Sites

Adler Planetarium & Astronomy Museum, Chicago
http://www.adlerplanetarium.org/

Astronomiae Historia/History of Astronomy
http://www.astro.uni-bonn.de/~pbrosche/astoria.html

Astronomy in Japan
http://www2.gol.com/users/stever/jastro.html

The Galileo Project
http://es.rice.edu/ES/humsoc/Galileo/

Institute and Museum of History of Science, Florence, Italy
http://galileo.imss.firenze.it/

Islamic Science
http://www.ou.edu/islamsci/

Map History/History of Cartography
http://ihr.sas.ac.uk/maps/

Museum of the History of Science, Oxford, England
http://www.mhs.ox.ac.uk/

List of Illustrations

All illustrations are from the historical collections of the Adler Planetarium & Astronomy Museum. They are reproduced with permission. Accession numbers are given in brackets at the end of relevant citations.

Index